Network Security

Network Security

André Perez

WILEY

First published 2014 in Great Britain and the United States by ISTE Ltd and John Wiley & Sons, Inc.

ISTE Ltd
27-37 St George's Road
London SW19 4EU
UK

www.iste.co.uk

John Wiley & Sons, Inc.
111 River Street
Hoboken, NJ 07030
USA

www.wiley.com

Library of Congress Control Number: 2014945531

British Library Cataloguing-in-Publication Data
A CIP record for this book is available from the British Library
ISBN 978-1-84821-758-4

Contents

Preface

This book introduces the security mechanisms deployed in
Ethernet, wireless-fidelity (Wi-Fi), Internet Protocol (IP) and
Multi-Protocol Label Switching (MPLS) networks. These
mechanisms are grouped according to the four functions
below:

– data protection;

– access control;

– network isolation;

– data monitoring.

Data protection is supplied by data confidentiality and
integrity control services:

– confidentiality consists of ensuring that data can only be
read by authorized individuals. This service is obtained
using a mechanism that encrypts the relevant data;

– integrity control consists of detecting modifications in
transferred data. This service is obtained via a hash function
or an encryption algorithm that generates a seal.

Access control is provided by a third-party authentication
service. This service consists of verifying the identity of the

person wishing to access a network. This service is generally obtained with a hash function, as for integrity control.

Network isolation is supplied by the Virtual Private Network (VPN) service. This service makes it possible to create closed user groups and authorize communication solely between users belonging to the same group. Note that access control also implicitly enables network isolation.

Data monitoring consists of applying rules to data in order to authorize its transfer or detect attacks. The service is supplied by analyzing the fields of the various protocols making up the data structure.

Network

The role of the network is to transport data between two hosts. The network is composed of two entities (Figure P.1):

– the Local Area Network (LAN) is the network on which the hosts connect. This is usually a private network deployed by businesses;

– the Wide Area Network (WAN) is the network that ensures the interconnection of the LAN networks. It is usually a public network deployed by Internet access and transit operators.

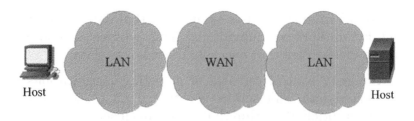

Figure P.1. *Network architecture*

The LAN network is constructed of two types of blocks: the access block and the core block (Figure P.2):

– the access block connects the network's hosts. Access blocks can be dedicated to different types of hosts:

- computers, telephones,

- application servers,

- network and security management system,

- WAN network;

– the core block enables the networking of access blocks.

The Internet access provider's WAN network is structured in three units (Figure P.3):

– access network: it corresponds to the connection of the LAN network with the operator's primary technical site;

– aggregation network: it collects the traffic generated by access networks. It generally has regional coverage;

– core network: this connects the different aggregation networks. It generally has national coverage and also provides the interface between operators.

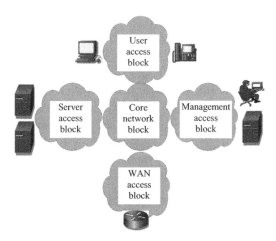

Figure P.2. *Architecture of LAN network*

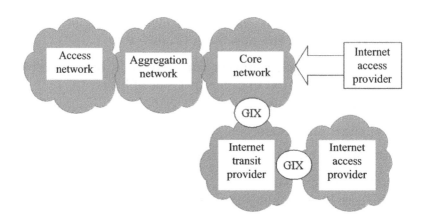

Figure P.3. *Architecture of WAN network*

The interconnection of the WAN networks of different Internet access providers takes place in the core network, in two different ways:

– the connection is direct, when the Internet access providers are operating in the same territory;

– the connection is established by an Internet transit provider, in the opposite case. Internet transit networks have an architecture similar to that of the core network of the Internet access provider.

A Global Internet eXchange (GIX) point enables different Internet access and transit providers to exchange traffic using mutual agreements called peering, generally based on the balancing of the volumes of data transmitted and received (Figure P.3).

Introduction to cryptography

Chapter 1 introduces the fundamental concepts of cryptography. Cryptography addresses aspects related to communications security, in order to provide confidentiality, integrity control and third-party authentication services.

Confidentiality service is implemented by encryption mechanisms. There are two main families of cryptographic algorithms: secret-key symmetrical algorithms and public- and private-key asymmetrical algorithms.

Symmetrical algorithms are well adapted to data encryption but pose the issue of establishing a secret key. Two frequently used methods are generation using the Diffie–Hellman algorithm and transport of the secret key via asymmetrical algorithms. Encryption is provided, for example, by the Advanced Encryption Standard (AES) algorithm or the Triple Data Encryption Standard (3DES) algorithm.

Asymmetrical algorithms are applied in the transportation of secret keys and digital signatures. In the former case, the secret key is encrypted by the public key and decrypted by the private key. In the latter case, the inverse occurs. Encryption is provided by algorithms based on modular exponentiation, such as the RSA (named after the initials of its three inventors, Rivest, Shamir and Adleman) algorithm or the Elliptic Curve Cryptography (ECC) algorithm.

The hash function is another type of cryptographic function. It converts a string of any length (the data to be protected) into a smaller chain, generally of fixed size (a digest). The hash function can be supplied by the two algorithms below:

– Message Digest 5 (MD5), which calculates a 128-bit digest;

– Secure Hash Algorithm (SHA), which calculates a digest of between 160 and 512 bits.

Sealing is based on the secret key and provides the data integrity control service. Seals can be calculated in two different ways:

– the symmetrical encryption algorithm is applied to the data; therefore, the seal is the last block of the cryptogram;

– the hash function is applied to a set comprised of data and a secret key, the association of which is defined, for example, by the Hashed Message Authentication Code (HMAC) calculation function.

The signature is based on the encryption of the digest by a private key and its decryption by a public key. It provides the service of integrity and non-repudiation by the source of the data received by the recipient.

The distribution of public keys is associated with the display of a certificate. This is a data structure signed by a certification authority that guarantees that the issuer of the public key is actually its holder.

802.1x mechanism

Chapter 2 introduces the 802.1x access control mechanism, deployed in the LAN network and implementing the following technologies:

– Ethernet technology, in the case of access to a switch;

– Wi-Fi technology, in the case of connection to an Access Point (AP).

The 802.1x mechanism has three components (Figure P.4):

– the supplicant is the device (for example, the computer) wishing to access the Ethernet or Wi-Fi network;

– the authenticator is the device (Ethernet switch or Wi-Fi AP) that controls the supplicant's access to the LAN network;

– the authentication server is the device that authenticates the supplicant and authorizes access to the LAN network.

The 802.1x mechanism is based on a series of protocols (Figure P.4):

– the Extensible Authentication Protocol (EAP) Over LAN (EAPOL) protocol, exchanged between the supplicant and the authenticator;

– the EAP protocols exchanged between the supplicant on one hand, and the authenticator or authentication server on the other. The EAP protocol is carried by the EAPOL protocol on the interface between the supplicant and the authenticator;

– the Remote Authentication Dial-In User Service (RADIUS) protocol exchanged between the authenticator and the authentication server. The RADIUS protocol carries the EAP protocol on the interface between the authenticator and the authentication server;

– the EAP-Method protocol exchanged between the supplicant and the application server. The EAP-Method protocol is carried by the EAP protocol.

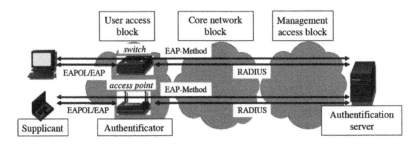

Figure P.4. *802.1x mechanism*

The EAP-Method protocol offers several types of authentication:

– the EAP-MD5 method: the client is authenticated using a password. This method is similar to the

Challenge-Handshake Authentication Protocol (CHAP), which is based on the Point-to-Point Protocol (PPP) used for point-to-point connections;

– the EAP-Transport Layer Security (TLS) method: authentication is mutual between the supplicant and the authentication server using certificates;

– the EAP-Tunneled-TLS (TTLS) method: authentication is mutual between the supplicant and the authentication server by means of an authentication server-side certificate, while the supplicant can use a password.

As a complement to authentication, the EAPOL protocol participates in the generation of keys for encryption and in the integrity control used by the Wi-Fi Protected Access (WPA) and WPA2 mechanisms described in Chapter 3.

WPA mechanisms

Chapter 3 introduces the WPA1 and WPA2 security mechanisms applied to Wi-Fi networks. Wi-Fi technology is originally a LAN private network access technology using radio transmission. It has the distinctive characteristic of using free frequency bands. It is also used in WAN public networks to establish hotspots.

WPA1 and WPA2 security mechanisms are used only in private networks. The security used in the case of hotspots generally involves the transport security described in Chapter 5.

Radio interface security began with the Wired Equivalent Privacy (WEP) mechanism. Due to its weaknesses, it was supplanted by the WPA1 mechanism and then by the WPA2 mechanism. These three mechanisms specifically implement third-party access control and data protection services.

For the WEP mechanism, third-party access control is based on the rivest cipher 4(RC4) algorithm. Access control takes place during the authentication phase, which is a procedure associated with the Medium Access Control (MAC) data connection protocol.

WPA1 and WPA2 mechanisms use the 802.1x mechanism, described in Chapter 2, for access control. The authentication phase is preceded by the security policy agreement procedure between the access point and the station.

For WEP and WPA1 mechanisms, encryption is performed by the RC4 algorithm. For the WEP mechanism, the master key is used for encrypting each Wi-Fi frame.

For the WPA1 mechanism, encryption is obtained using a key derived from the master key for temporary use. In association with encryption, a protocol containing the initialization vector is added to the MAC data connection protocol (Figure P.5):

– the WEP protocol in the case of the WEP mechanism;

– the Temporal Key Integrity Protocol (TKIP) in the case of the WPA1 mechanism.

For the WPA2 mechanism, encryption is based on the AES algorithm and the header of the MAC data connection protocol is completed by the Counter-mode/Cipher block chaining MAC Protocol (CCMP) header (Figure P.5).

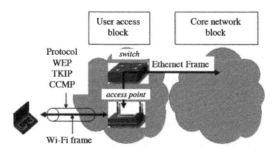

Figure P.5. *WEP, TKIP and CCMP protocols*

Integrity control is provided by a cyclic redundancy check (CRC) for the WEP mechanism. The WPA1 mechanism uses the Michael algorithm. In the case of the WPA2 mechanism, integrity control is obtained using the AES algorithm.

IPSec mechanism

Chapter 4 introduces the Internet Protocol Security (IPSec) mechanism applied to the IP packet. There are two parts to this mechanism (Figure P.6):

– the establishment of the security association between two security gateways, located in the LAN network and the WAN network access block;

– the protection of data between these two gateways.

Security association takes place in two phases:

– phase one consists of authenticating the security gateways wishing to establish the security association;

– phase two makes it possible to establish the parameters to be used for implementing data protection (protocol, algorithm and key).

Two versions of protocols have been defined for the establishment of the security association. The first version has three parts:

– the Internet Security Association and Key Management Protocol (ISAKMP) defines the framework of the establishment, modification and deletion of the security association;

– the Domain of Interpretation (DOI) document defines the parameters negotiated relative to the use of the ISAKMP;

– the internet key exchange (IKEv1) mechanism defines the exchange procedures relevant to the use of the ISAKMP.

Chapter 4 describes only the second IKEv2 version, which simplifies the previous version. This version combines the functionalities defined in IKEv1 and ISAKMP, deleting their unnecessary functions. It eliminates the generic character of the previous version by integrating the DOI function, which defines the parameters specific to the security association.

Data protection introduces two extensions of the IPv4 or IPv6 header:

– the Authentication Header (AH) is designed to ensure integrity control without data encryption (without confidentiality);

– the Encapsulating Security Payload (ESP) ensures integrity control and confidentiality of IP packets.

The protection of data between two security gateways uses the tunnel mode. This mode is characterized by the fact that the AH or ESP header encapsulates the source IP packet, and the whole is then encapsulated in turn by a new IP header.

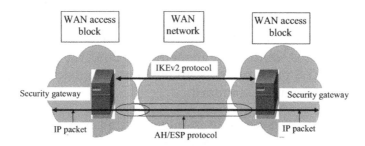

Figure P.6. *IPSec mechanism*

SSL / TLS / DTLS protocols

Chapter 5 introduces Secure Sockets Layer (SSL) and Transport Layer Security (TLS) protocols pertaining to the transport of data, applied To Transmission Control Protocol

(TCP) segments. The Datagram TLS (DTLS) protocol is an adaptation for the User Datagram Protocol (UDP), Datagram Congestion Control Protocol (DCCP), Stream Control Transmission Protocol (SCTP) and Secure Real-time Transport Protocol (SRTP).

The TLS protocol is standardized by the Internet Engineering Task Force (IETF). It is the successor of the SSL protocol, developed by Netscape, the original purpose of which was to establish the security of exchanges between a navigator and a web server.

Several versions of the TLS protocol have since been defined: TLS 1.0, TLS 1.1, and TLS 1.2. Version TLS 1.0 corresponds to SSL version 3.1, which is the latest version of the SSL protocol. The differences between SSL 3.0, present in navigators, and TLS 1.0 are minimal, but still sufficient to make these protocols incompatible.

TLS 1.0 has made it possible to correct a cryptographic flaw in SSL 3.0 and propose cryptographic algorithms concerning key exchange and authentication. TLS 1.1 is a version that enables protection against attacks shown on the use of Cipher Block Chaining (CBC) -mode encryption. TLS 1.2 integrates scattered elements within the norm and describes TLS extensions as an element complete in its own right apart from the standard.

Data transport security is implemented between a client who initializes the session and a security gateway acting as a server, located within the LAN network in the WAN network access block.

SSL/TLS protocols correspond to a *Record* header that encapsulates SSL/TLS messages or application-layer data and SSL/TLS exchanges between the client and the security gateway:

– the *change_cipher_spec* message indicates a modification of security parameters;

– the *alert* message indicates an error in communication between the host and the security gateway;

– *handshake* messages negotiate the security parameters between the host and the security gateway.

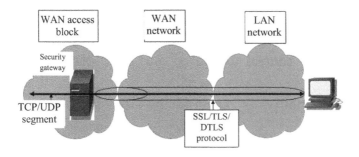

Figure P.7. *SSL, TLS and DTLS protocols*

Network management

Chapter 6 introduces security mechanisms pertaining to protocols related to network management.

The Simple Network Management Protocol (SNMP) enables the execution of equipment management functions (a switch or router) divided into three categories (Figure P.8):

– supervision or alarm management;

– configuration management;

– performance management.

The protocol SNMPv1 is the first version of the protocol. Security is based on a string of characters called a community, which supply read-only rights or read–write rights. This version has the disadvantage of transporting this password unscrambled in the SNMP message.

The protocol SNMPv2 is the second version of the protocol. It completes the structure of the management information base (MIB) database describing the equipment in the form of objects. The protocol is also enriched by new messages. However, no changes were made to the security of exchanges.

The protocol SNMPv3 is the third version of the protocol. Its main contribution lies in the introduction of stronger security mechanisms:

– integrity control is based on the MD5 or SHA-1 algorithm;

– confidentiality is ensured by the DES encryption algorithm.

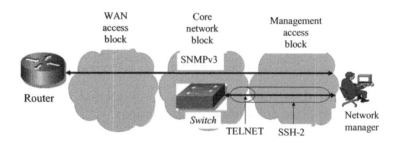

Figure P.8. *Network administration*

Telnet is a protocol that enables a direct link between a client located on the administrator platform side and a command interpreter on the server side, localized in the equipment to be administrated. Telnet sessions are opened on the basis of a password circulating unscrambled between the client and the server.

The Secure SHell (SSH) protocol provides authentication, integrity control and confidentiality services for the Telnet messages exchanged (Figure P.8). SSH-2 is the standardized version of the protocol. It is composed of three parts:

– SSH Transport Layer Protocol (SSH-TRANS) is the protocol supplying the integrity control and confidentiality bases;

– SSH Authentication Protocol (SSH-AUTH) is the protocol enabling client authentication;

– SSH Connection Protocol (SSH-CONN) is the protocol enabling several sessions to be maintained on one SSH connection.

MPLS technology

Chapter 7 introduces the mechanisms enabling the isolation of IP packets in a WAN network.

WAN-network access and aggregation networks are actually Ethernet networks, the isolation of which is described in Chapter 8 (Figure P.9).

The core network of the WAN network is an MPLS network integrating the IP VPN function (Figure P.9).

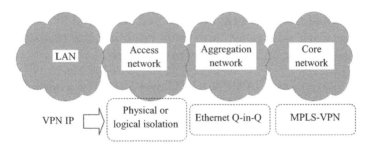

Figure P.9. *IP VPN*

The MPLS function consists of labeling an IP packet and using this label for replacement switching of IP routing. This Label Switching Path (LSP) label is carried by an MPLS header inserted between layer 3 (IP) and layer 2 (usually Ethernet).

The Label Forwarding Information Base (LFIB) table is implemented via two protocols:

– the Label Distribution Protocol (LDP), which associates a label with a network IP address. This protocol feeds the Label Information Base (LIB).

– the Open Shortest Path First (OSPF) or Intermediate System to Intermediate System (IS-IS) routing protocol, which determines an exit port for an IP network address. This table informs the Routing Information Base (RIB).

The VPN function is executed by Provider Edge (PE) devices surrounding the core network. It consists of introducing the following mechanisms:

– isolation of the routing table in order to propagate routes only in cases of routing specific to a closed user group;

– marking of IP packets with a particular label. This VPN label is carried by an additional MPLS header.

The MPLS-VPN network introduces the public VPN-IPv4 addresses of the network, constructed using public or private IPv4 network addresses, and thus enabling the creation of a private network on a public infrastructure.

The distribution of VPN labels and VPN-IPv4 addresses is ensured by the multiprotocol border gateway protocol (MP-BGP-4) routing protocol exchanged between the surrounding PE devices.

The MPLS-VPN network also enables the construction of complex VPN architectures from the import and export rules of IPv4 routes.

Ethernet VPN

Chapter 8 introduces the mechanisms enabling Ethernet frames to be isolated in LAN and WAN networks.

The isolation of Ethernet frames in a LAN network is executed by the Virtual LAN (VLAN) or Q-VLAN function. It is based on the marking of Ethernet frames, with each mark corresponding to a closed user group.

Ethernet frames may be isolated in a WAN network using the three technologies below:

– Provider Bridge Transport (PBT);

– Virtual Private LAN Service (VPLS);

– Layer 2 Tunneling Protocol (L2TPv3);

PBT technology can be considered as an extension of the Q-VLAN function executed in the LAN network. It is generally deployed in the aggregation network via the implementation of double-marking (Q-in-Q) of Ethernet frames.

PBT technology has also defined the isolation of Ethernet frames in the core network via the implementation of a double Ethernet header (MAC-in-MAC). This function is not deployed and, therefore, is not described in Chapter 8.

VPLS technology is an extension of MPLS technology. It is deployed in the core network and has the advantage of sharing provider (P) equipment with the MPLS-VPN network.

VPLS technology can be expanded to the aggregation network with the Hierarchical VPLS (H-VPLS) function.

L2TPv3 technology is a functionality implemented for the transfer of Ethernet frames, solely point-to-point, via a network of IP routers.

In summary, the implementation of Ethernet VPN in the WAN network is obtained in various ways (Figure P.10):

– In the access network, isolation is physical or logical depending on the type of technology used. The connection of two users in the access network is generally prohibited. User-generated traffic is required to be transmitted toward the aggregation network.

– In the aggregation network, isolation is executed using the H-VPLS function or Ethernet double marking (Q-in-Q);

– In the core network, isolation is executed by the VPLS, MAC-in-MAC or L2TPv3 functions.

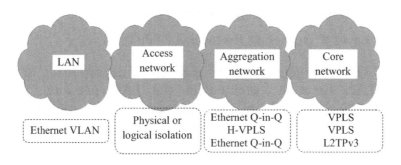

Figure P.10. *Ethernet VPN*

Firewalls

Chapter 9 introduces the functionalities of firewalls. They ensure the surveillance of data transiting between LAN and WAN networks by controlling the fields of different protocols according to established rules.

There are several types of firewalls, which are characterized by the following functions:

– stateless packet filtering: this control is applied to fields in IP, TCP, UDP and Internet Control Message Protocol (ICMP) headers;

– stateful inspection implements controls of the TCP state machine. Controls are conducted on sequences of TCP segments.

– application message filtering: this function is carried out by Application-Layer Gateway (ALG), which inspects message content.

Firewalls are deployed in the LAN network within the WAN-network access block. They are integrated into the DeMilitarized Zone (DMZ).

Two packet filters surround the DMZ:

– one of the packet filters (*front-end firewall*) inspects packets exchanged between the WAN network and the demilitarized zone;

– the other packet filter (*back-end firewall*) inspects packets exchanged between the LAN network and the demilitarized zone.

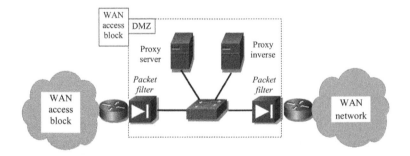

Figure P.11. *Demilitarized zone*

The DMZ hosts two types of application firewalls: the proxy server and the inverse proxy server. The proxy server (inverse proxy, respectively) executes a flow control stemming from the client (toward the server, respectively) connected to the LAN network.

Network Address Translation (NAT), or Network Address and Port Translation (NAPT), is a specific type of firewall. This function authorizes only traffic initialized by LAN network hosts and blocks any attempt at connection coming from the WAN network.

Several NAPT configurations define a more or less selective filter: open cone, address-restricted cone, port-restricted cone and symmetrical cone.

The NAT/NAPT device presents difficulties for some flows crossing it, for which specific mechanisms are defined:

– applications with a specific identification such as the ICMP;

– protocols protecting the IP packet load, such as for example the ESP protocol of the IPSec security mechanism.

– applications transporting IP addresses, such as Session Information Protocols (SIP) and Session Description Protocols (SDP);

– flows established dynamically, such as File Transfer Protocols (FTP) or Real-time Transport Protocols (RTP);

– fragmented IP packets.

Intrusion detection

Chapter 10 introduces the functionalities of Intrusion Detection Systems (IDS) and Intrusion Prevention Systems (IPS). These two types of systems are grouped under the term Intrusion Detection Prevention System (IDPS).

These devices ensure the monitoring of data transiting between LAN and WAN networks, or within the LAN network, via data analysis enabling the detection of attacks.

Intrusion detection methods are implemented via the following techniques:

– detection based on the signatures of known attacks in data circulating within the LAN network;

– detection of anomalies via an analysis of suspicious activities in the behavior of a host;

– analysis of protocols in order to verify their compliance with norms.

Different types of IDPS devices are deployed in the network according to the localization or function fulfilled:

– Network-based IDPS (N-IDPS): this device enables the monitoring of data on various segments of the LAN. This device is generally applied to Ethernet interfaces;

– Wireless IDPS (WIDPS): this device enables the monitoring of data transiting via Wi-Fi radio interface. It constitutes a specific case of the N-IDPS device;

– Home-based IDPS (H-IDPS): this device presents functionalities similar to the devices above. It enables the monitoring of data solely in the network hosts;

– Network behavior analysis (NBA): this device enables the specific execution of traffic analysis in order to detect unusual activity.

Abbreviations

3DES	Triple Data Encryption Standard
AAD	Additional Authentication Data
AC	Attachment Circuit
ACL	Access Control List
AES	Advanced Encryption Standard
AH	Authentication Header
AKM	Authentication and Key Management
ALG	Application-Layer Gateway
AP	Access Point
ARP	Address Resolution Protocol
AS	Autonomous System
ASN.1	Abstract Syntax Notation
ATM	Asynchronous Transfer Mode
AVP	Attribute Value Pair
BSS	Basic Service Set
BSSID	BSS Identifier
CCMP	Counter-mode/Cipher block chaining MAC Protocol
CDN	Circuit-Disconnect-Notify
CE	Customer Edge
CFI	Canonical Format Indicator

CHAP	Challenge-Handshake Authentication Protocol
CRC	Cyclic Redundancy Check
C-TAG	Customer TAG
DCCP	Datagram Congestion Control Protocol
DEI	Drop Eligible Indicator
DES	Data Encryption Standard
DF	Don't Fragment
DIX	DEC, Intel et Xerox
DLCI	Data Link Connection Identifier
DMZ	DeMilitarized Zone
DNS	Domain Name System
DOI	Domain of Interpretation
DoS	Denial of Service
DSAP	Destination Service Access Point
DSCP	DiffServ Code Point
DTLS	Datagram TLS
EAP	Extensible Authentication Protocol
EAPOL	EAP Over LAN
ECC	Elliptic Curve Cryptography
ECN	Explicit Congestion Notification
E-LSP	EXP-inferred-class LSP
EoMPLS	Ethernet over MPLS
ESP	Encapsulating Security Payload
ESS	Extented Service Set
FEC	Forwarding Equivalent Classes
FTP	File Transfer Protocol
GEK	Group Encryption Key
GIK	Group Integrity Key
GIX	Global Internet eXchange
GMK	Group Master Key
GTK	Group Transient Key

GTKSA	GTK Security Association
H-IDPS	Home-based IDPS
HMAC	Hashed Message Authentication Code
ICCN	Incoming-Call-Connected
ICE	Interactive Connectivity Establishment
ICMP	Internet Control Message Protocol
ICRP	Incoming-Call-Reply
ICRQ	Incoming-Call-Request
ICV	Integrity Check Value
IDPS	Intrusion Detection Prevention System
IDS	Intrusion Detection System
IE	Information Element
IEEE	Institute of Electrical and Electronics Engineers
IETF	Internet Engineering Task Force
IHL	Internet Header Length
IKE	Internet Key Exchange
IP	Internet Protocol
IPS	Intrusion Prevention System
IPSec	Internet Protocol Security
ISAKMP	Internet Security Association and Key Management Protocol
IS-IS	Intermediate System to Intermediate System
ISP	Internet Service Protocol
IV	Initialization Vector
KCK	Key Confirmation Key
KEK	Key Encryption Key
L2TPv3	Layer 2 Tunneling Protocol
LAC	L2TP Access Concentrator
LAN	Local Area Network
LDP	Label Distribution Protocol
LFIB	Label Forwarding Information Base

LIB	Label Information Base
LLC	Logical Link Control
L-LSP	Label-inferred-class LSP
LNS	L2TP Network Server
LSP	Label Switching Path
LSR	Label Switching Router
MAC	Medium Access Control
MAC	Message Authentication Code
MD5	Message Digest 5
MF	More Fragment
MIB	Management Information Base
MIC	Message Integrity Code
MK	Master Key
MODP	MODular exponential modulus P
MP-BGP-4	Multi-Protocol - Border Gateway Protocol 4
MPLS	Multi-Protocol Label Switching
MSDU	MAC Service Data Unit
MTU	Maximum Transmission Unit
NAPT	Network Address and Port Translation
NAT	Network Address Translation
NAT-D	NAT Discovery
NAT-OA	NAT Original Address
NAT-T	NAT Transversal
NBA	Network Behavior Analysis
N-IDPS	Network-based IDPS
NIST	National Institute of Standards and Technology
N-PE	Network-facing PE
OSA	Open System Authentication
OSPF	Open Shortest Path First
OUI	Organizationally Unique Identifier
P	Provider

PAD	Peer Authorization Database
PB	Provider Bridge
PBT	Provider Bridge Transport
PCP	Priority Code Point
PDU	Protocol Data Unit
PE	Provider Edge
PHB	Per-Hop Behavior
PHP	Penultimate Hop Popping
PKI	Public Key Infrastructure
PMK	Pairwise Master Key
PMTU	Path MTU
PN	Packet Number
PPP	Point-to-Point Protocol
PSK	Pre-Shared Key
PTK	Pairwise Transient Key
PTKSA	PTK Security Association
PW	Pseudo-Wire
RADIUS	Remote Authentication Dial-In User Service
RC4	Rivest Cipher 4
RD	Route Distinguisher
RFC	Request For Comments
RIB	Routing Information Base
SKA	Shared Key Authentication
RSA	Rivest, Shamir, Adleman
RSN	Robust Security Network
RSVP	ReSerVation Protocol
RT	Route Target
RTCP	RTP Control Protocol
RTP	Real-time Transport Protocol
SA	Security Association
SAD	Security Association Database

SCCCN	Start-Control-Connection-Connected
SCCRP	Start-Control-Connection-Reply
SCCRQ	Start-Control-Connection-Request
SCTP	Stream Control Transmission Protocol
SDP	Session Description Protocol
SFD	Start of Frame Delimiter
SHA	Secure Hash Algorithm
SIP	Session Information Protocol
SMI	Structure of Management Information
SNAP	Sub-Network Access Protocol
SNMP	Simple Network Management Protocol
SPD	Security Policy Database
SPI	Security Parameters Index
SRTP	Secure Real-time Transport Protocol
SSAP	Source Service Access Point
SSH	Secure Shell
SSID	Service Set Identifier
SSL	Secure Sockets Layer
S-TAG	Service TAG
StopCCN	Stop-Control-Connection-Notification
STP	Spanning Tree Protocol
STUN	Session Traversal Utilities for NAT
TCI	Tag Control Identification
TCP	Transmission Control Protocol
TK	Temporary Key
TKIP	Temporal Key Integrity Protocol
TL	Tunnel Label
TLS	Transport Layer Security
TLV	Type Length Value
TMK	Temporary MIC Key
ToS	Type of Service

TPID	Tag Protocol Identification
TSC1	TKIP Sequence Counter
TTL	Time to Live
TTLS	Tunneled- TLS
TURN	Traversal Using Relay NAT
UDP	User Datagram Protocol
U-PE	User-facing PE
USM	User-based Security Model
UTM	Unified Threat Management
VACM	View-Based Access Control Model
VCI	Virtual Channel Identifier
VCL	Virtual Channel Label
VID	VLAN Identification
VLAN	Virtual LAN
VPI	Virtual Path Identifier
VPLS	Virtual Private LAN Service
VPN	Virtual Private Network
VRF	VPN Routing and Forwarding
VSI	Virtual Switch Instance
WAN	Wide Area Network
WEP	Wired Equivalent Privacy
WIDPS	Wireless IDPS
Wi-Fi	Wireless-Fidelity
WLAN	Wireless LAN
WPA	Wi-Fi Protected Access
XOR	eXclusive OR

1

Introduction to Cryptography

1.1. The encryption function

The encryption function is the mechanism used to provide a confidentiality service. It enables the modification of a string of bytes (the data being transmitted) in order to make it incomprehensible to anyone who is not authorized to know its content.

Encryption is done using two types of algorithm:

– symmetrical or secret-key algorithms. The same (secret) key is used for encryption and decryption;

– asymmetrical or public key algorithms. Different keys are used for encryption and decryption. The public key (or, conversely, the private key) is used for encryption. The private key (or, conversely, the public key) is used for decryption.

Symmetrical algorithms are grouped into two categories:

– stream cipher algorithms act on bits. A stream cipher generally consists of an exclusive OR or XOR (eXclusive OR) operation between data issued by a pseudo-random number-generator and the data being transmitted;

– block cipher algorithms act on blocks of between 32 and 512 bits in size.

Asymmetrical algorithms are based on modular exponentiation or an elliptical curve.

Asymmetrical algorithms based on modular exponentiation are not adapted to data encryption because they have the disadvantage of being slow. This is due to the size of the keys used. They are used mainly in the following two scenarios:

– secret-key transport. The secret key is encrypted by the public key and decrypted by the private key. Only the holder of the private key can recover the secret key, which guarantees confidentiality;

– signature. The data digest, calculated using a hash function, is encrypted by the private key and decrypted by the public key. The data source is the only holder of the private key, which guarantees the integrity control of the data received, the authentication of the data source, and the non-repudiation by the client.

Table 1.1 shows a comparison of key sizes depending on whether symmetrical and asymmetrical algorithms are used to obtain an equivalent level of security.

Symmetrical algorithm	Asymmetrical algorithm Modular exponentiation	Asymmetrical algorithm Elliptical curve
80	1,024	160
112	2,048	224
128	3,072	256
192	7,680	384
256	15,360	512

Table 1.1. *Comparison of key size depending on algorithm*

1.1.1. *3DES algorithm*

The triple data encryption standard (3DES) algorithm is a symmetrical algorithm by blocks. It strings three successive operations of the DES algorithm on a single 64-bit data block.

Three keys (key1, key2 and key3) are used for encryption, one key per operation. Each 64-bit key contains 56 randomly generated bits and 8 odd parity check bits. Two options are defined for the composition of keys:

– option 1: the three keys (key1, key2 and key3) are different;

– option 2: two keys (key1 and key2) are different; two keys (key1 and key3) are identical.

The 3DES algorithm encryption operation consists of sequencing a DES encryption (E) with the key1, a DES decryption (D) with the key2, and a DES encryption (E) with the key3.

Output data = $E_{key3}(D_{key2}(E_{key1}(\text{input data})))$.

The DES encryption algorithm is shown in Figure 1.1.

Each 64-bit data block is submitted to an initial permutation function. The resulting 64-bit block is then cut into two 32-bit blocks (left block L_0 and right block R_0).

The DES algorithm is constructed using 16 successive iterations. The first iteration generates, from the two blocks L_0 and R_0, two blocks L_1 and R_1 in the following manner:

$L_1 = R_0$

$R_1 = L_0 \oplus f(R_0, K_1)$

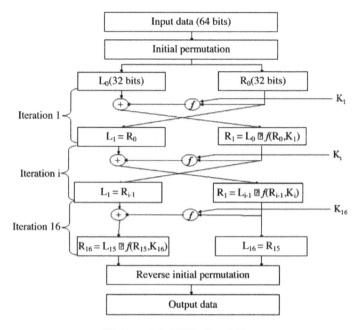

Figure 1.1. *DES algorithm*

At each iteration, the same operation is executed:

– left block L_i corresponds to right block R_{i-1};

– right block R_i is an exclusive OR of left block L_{i-1} and of the result of an *f* function applied to right block R_{i-1} and to K_i deducted from the initial 64-bit key (key1, key2 or key3).

The output of the last iteration is composed of left block L_{16} and right block R_{16}. Block $R_{16}L_{16}$ is subjected to a reverse initial permutation.

The *f* function is described in Figure 1.2.

The E function is used to generate 48 bits from right block R. The exclusive OR operation is applied to the previous result and to key K.

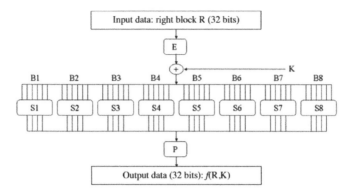

Figure 1.2. *f function*

The 48 resulting bits are cut into eight blocks (B1 to B8) of six bits, to which substitution functions are applied (S1 to S8), resulting in an output of a four-bit block for each substitution function.

The permutation function P operates on the 32 bits produced by the eight substitution functions S1 to S8.

The *f* function is written as follows:

$f(K,R) = P(S1(B1)S2(B2)...S8(B8))$, with B1B2...B8 = K $\oplus$ E(R)

The generation of keys K_i is shown in Figure 1.3.

A permutation function P1 is applied to the 64 bits of the initial key (key1, key2, or key3). The resulting 56 bits are cut into two blocks C_0 and D_0.

Blocks C_i and D_i are deduced from the previous blocks C_{i-1} and D_{i-1} by creating a shift to the left of one or two bits depending on the number of the block.

A P2 permutation is applied to this C_iD_i block to obtain the 16 48-bit keys K_i used for the 16 iterations.

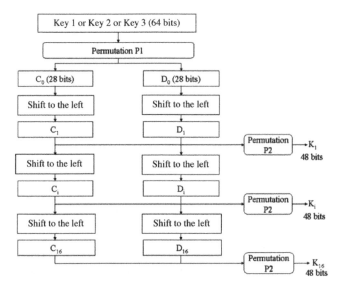

Figure 1.3. *Generation of keys K_i*

1.1.2. *AES algorithm*

The advanced encryption standard (AES) algorithm is a symmetrical block algorithm. It operates on 128-bit data blocks and uses keys 128 bits, 192 bits, or 256 bits in length.

The 16-byte data input block is represented by a four-line by four-column table, with each cell containing a byte.

Input data: 32 43 f6 a8 88 5a 30 8d 31 31 98 a2 e0 37 07 34

32	88	31	e0
43	5a	31	37
f6	30	98	07
a8	8d	a2	34

The 16-byte key is represented by a four-line by four-column table, with each cell containing a byte.

Initial 128-bit key: 2b 7e 15 16 28 ae d2 a6 ab f7 15 88 09 cf 4f 3c

2b	28	ab	09
7e	ae	f7	cf
15	d2	15	4f
16	a6	88	3c

An exclusive OR is executed between the input data and the initial key. The output data makes up the entry data of the first iteration.

Input data					Key					Output data			
32	88	31	e0		2b	28	ab	09		19	a0	9a	e9
43	5a	31	37	⊕	7e	ae	f7	cf	=	3d	f4	c6	f8
f6	30	98	07		15	d2	15	4f		e3	e2	8d	48
a8	8d	a2	34		16	a6	88	3c		be	2b	2a	08

The SubBytes transformation applies the S-box substitution table to each byte of the input data table of the iteration. For example, if the byte has a hexadecimal value of 19, the substitution value corresponds to the value of the first line and ninth column in the S-box table.

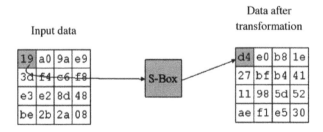

Input data

Data after transformation

Iteration 1: transformation SubBytes

The ShiftRows transformation consists of applying a shift to each line for the three last lines of the table resulting from the SubBytes transformation, with the first line being completely copied without modification. The second line undergoes a shift of one byte on the left; the third line a shift of two bytes; and the fourth line a shift of three bytes.

Data after transformation SubBytes

d4	e0	b8	1e
27	bf	b4	41
11	98	5d	52
ae	f1	e5	30

Shift of one byte on the left
Shift of 2 bytes on the left
Shift of 3 bytes on the left

Data after transformation ShiftRows

d4	e0	b8	1e
bf	b4	41	27
5d	52	11	98
30	ae	f1	e5

Iteration 1: transformation ShiftRows

The MixColumns transformation operates on the columns of the table resulting from the ShiftRows transformation.

The transformation of the first column is executed as follows:

$\{04\} = (\{02\} \bullet \{d4\}) \oplus (\{03\} \bullet \{bf\}) \oplus \{5d\} \oplus \{30\}$

$\{66\} = \{d4\} \oplus (\{02\} \bullet \{bf\}) \oplus (\{03\} \bullet \{5d\}) \oplus \{30\}$

$\{81\} = \{d4\} \oplus \{bf\} \oplus (\{02\} \bullet \{5d\}) \oplus (\{03\} \bullet \{30\})$

$\{e5\} = (\{03\} \bullet \{d4\}) \oplus \{bf\} \oplus \{5d\} \oplus (\{02\} \bullet \{30\})$

The symbol $\bullet$ represents a multiplication of bytes in the form of polynomials, modulo the first polynomial $x^8 + x^4 + x^3 + x + 1$.

Data after transformation ShiftRows

d4	e0	b8	1e
bf	b4	41	27
5d	52	11	98
30	ae	f1	e5

Transformation matrix

```
02 03 01 01
01 02 03 01
01 01 02 03
03 01 01 02
```

Data after transformation MixColumns

04	e0	48	28
66	cb	f8	06
81	19	d3	26
e5	9a	7a	4c

Iteration 1: transformation MixColumns

The first iteration is ended by an exclusive OR (AddRoundKey transformation) between the output data of the `MixColumns` transformation and the key of the first iteration. This key is derived from the initial key.

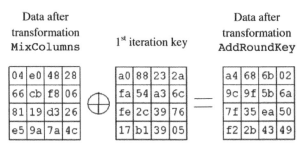

Data after transformation MixColumns	1st iteration key	Data after transformation AddRoundKey

Iteration 1: transformation AddRoundKey

The output data of the first iteration constitutes the input data of the second iteration. The number of iterations is equal to 10 for a 128-bit key, to 12 for a 192-bit key, and to 14 for a 256-bit key.

Table 1.2 shows the derivation of the initial key for the first iteration.

Key derivation starts with the cutting of the initial 128-bit key into four blocks, w_0, w_1, w_2 and w_3.

$w_0 = 2b7e1516$

$w_1 = 28aed2a6$

$w_2 = abf71588$

$w_3 = 09cf4f3c$

The key of the first iteration is made up of four blocks, w_4, w_5, w_6 and w_7. The `RotWord` transformation consists of the shift of a byte on the left. The `SubWord` transformation applies the S-box substitution table to each byte. Block w_i is the result of an exclusive OR between blocks w_{i-1} (having possibly undergone transformations) and w_{i-4}.

i	4	5	6	7
w_{i-1}	09cf4f3c	a0fafe17	88542cb1	23a33939
After RotWord	cf4f3c09			
After SubWord	8a84eb01			
Rcon	01000000			
After exclusive OU	8b84eb01			
w_{i-4}	2b7e1516	28aed2a6	abf71588	09cf4f3c
w_i	a0fafe17	88542cb1	23a33939	2a6c7605

Table 1.2. *Derivation of the initial key for the first iteration*

1.1.3. *RSA algorithm*

The RSA algorithm (named after the initials of its three inventors, Rivest, Shamir and Adleman) is a public key asymmetrical algorithm based on modular exponentiation.

Alice wants to receive a block of confidential data M from Bob. Alice generates a public key that she transmits to Bob and a private key that she keeps. Bob encrypts the data block M with the public key. Only Alice can recover the data block M using her private key.

The RSA algorithm is implemented using the following procedure:

– Alice generates two prime numbers, p and q: $p = 59$ and $q = 71$;

– Alice calculates the product $w = (p-1)(q-1)$: $w = 58 \times 70 = 4060$;

– Alice chooses a prime number e with w: $e = 671$. It is relatively easy to determine the value e using Euclid's algorithm to verify whether it is prime or not with w;

– Alice calculates the product $n = p \times q$: $n = 59 \times 71 = 4189$;

– Alice calculates the number d such that $d \times e \equiv 1\ [mod\ w]$: $d = 1791$. The calculation of the value d uses an extended Euclidean algorithm, or Bézout algorithm. Since e is prime with w, there are two integer numbers d and k such that $(e \times d) + (k \times w) = 1$ (Bachet–Bézout theorem);

– Alice transmits to Bob the values n and e, which constitute the public key, and keeps the values n and d, which represent the private key;

– Bob encrypts his message M with the public key: $M \mapsto M^{671}$ [mod 4189]M = 01010010 01010011 01000001 M is cut into blocks of length k such that $2^k < n < 2^{k+1}$ M is therefore cut into blocks of 12 bits: M = 010100100101 001101000001The 12 bits 010100100101 converted into decimals give 1317. The 12 bits 001101000001 converted into decimals give 833. The encrypted message C transmitted by Bob is as follows:

- Bob calculates 1317^{671} [mod 4189], or 3530,

- Bob calculates 833^{671} [mod 4189], or 3050;

– Alice decrypts the received message C with her private key: $C \mapsto C^{1791}$ [mod 4189]:

- Alice calculates 3530^{1791} [mod 4189] and recovers the value 1317,

- Alice calculates 3050^{1791} [mod 4189] and recovers the value 833.

The difficulty of the RSA algorithm lies in the determination by Alice of two prime numbers p and q and in the calculation of M^e [mod n] and C^d [mod n]. The generation of large prime numbers constitutes a relatively complex problem. There are various methods for the calculation of modular exponentiation.

1.1.4. *ECC algorithm*

The Elliptic Curve Cryptography (ECC) algorithm is a public key asymmetrical algorithm based on elliptic curves.

An elliptic curve is a set of points (x,y) satisfying the conditions below:

$$y^2 = x^3 + ax + b$$

with *4 a^3 + 27 b^2* different than zero

When E is an elliptic curve defined by values a and b, we can assign a switching operator + to this set. We define the sum of two points $(A, B) = A + B = C'$ on an elliptic curve as follows:

– a secant line passing through the two points A and B cuts the curve at a third point C;

– the point C' is the symmetric of C in relation to the axis of abscissas;

– if the two points A and B are identical, we consider the line connecting them is the tangent to the elliptic curve;

– if the line is parallel to the access of ordinates (the two points are P and $-P$), the point of intersection is considered as being infinite. This point is represented by point O such that $P + O = P$ or $P + -P = O$.

The procedure takes place as follows:

– Alice receives from Bob the public key composed of the values of a, b, P, and (d_bP):

- a and b are the values of the elliptic curve,

- P is a point on the elliptic curve,

- d_b is Bob's private key,

- (d_bP) is the point on the elliptic curve resulting from the sum of $d_b + P$;

– Alice chooses a secret number n;

– Alice sends Bob the two points nP and $[M + n(d_bP)]$, with M being the data block to be transmitted;

– Bob calculates $d_b(nP) = n(d_bP)$, then $[M + n(d_bP)] - n(d_bP) = M$, which enables him to recover the data block M.

1.2. Hash function

The hash function is the mechanism that converts a string of bytes (the data being transmitted) into a smaller string called a digest. When the data to be transmitted is associated with a secret key, the hash function enables the sealing of this data to provide authentication of the data source and integrity control of the data. The non-repudiation service is not provided.

1.2.1. *MD5 algorithm*

The Message Digest 5 (MD5) algorithm calculates a 128-bit digest from a block of data cut into 512-bit data blocks.

The first stage consists of filling the data block. The data block is completed so as to be congruent to 448 bits modulo 512. Filling is always executed, even if the message is already congruent to 448 bits modulo 512. The filling value therefore varies from 1 to 512 bits. Filling is executed as follows:

– the final filling bit is set at one;

– the other filling bits are set at zero.

The length of the data block is coded on 64 bits. This field is added to the structure previously defined so as to obtain 512-bit data blocks.

Calculation of the digest is based on registers A, B, C and D, which are 32 bits in length. For the first 512-bit block, these registers are initialized with the following values (initialization vector) in hexadecimals (Figure 1.4):

A = 67452301

B = EFCDAB89

C = 98BADCFE

D = 10325476

For the following 512-bit blocks, these registers are initialized from the values obtained for the previous 512-bit block. When all of the 512-bit blocks have been processed, the final digest corresponds to the concatenation of the values of registers A, B, C and D (Figure 1.4).

Operations on a 512-bit block are broken down into four stages, which are themselves subdivided into 16 primary iterations based on a function that varies according to stage, on an addition and a shift to the left. When the four stages are finished, the values of registers A, B, C and D are added to those of the initialization vector (Figure 1.4).

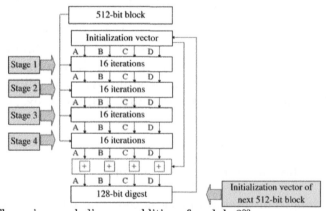

Note: The + sign symbolizes an addition of modulo 2^{32}

Figure 1.4. *MD5 algorithm*

The four functions defined are:

Stage one: F(X, Y, Z) = (X AND Y) OR ((NO X) AND Z)

Stage two: F(X, Y, Z) = (X AND Y) OR (Y AND (NO Z))

Stage three: F(X, Y, Z) = X XOR Y XOR Z

Stage four: F(X, Y, Z) = Y XOR (X OR (NO Z))

The primary iteration is shown in Figure 1.5. The new values of registers A, B, C and D are supplied, for stage one, by the following equations:

newA = previousD

newB = ((previousA + F(previousB, previousC, previousD) + k[i] + w[g]) <<< r[i]) + previousB.

newC = previousB

newD = previousC

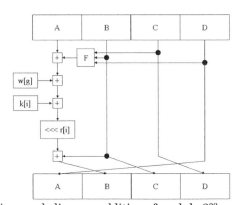

Note: The + sign symbolizes an addition of modulo 2^{32}

Figure 1.5. *Primary iteration of MD5 algorithm*

Each 512-bit data block is subdivided into 16 words w[i] of 32 bits with $0 \leq i \leq 15$. The rank of the stage determines the index value [g]:

g = i, with $0 \leq i \leq 15$ for stage one

g = (5×i + 1) [mod 16], with 16 ≤ i ≤ 31 for stage two

g = (3×i + 5) [mod 16], with 32 ≤ i ≤ 47 for stage three

g = (7×i) [mod 16], with 48 ≤ i ≤ 63 for stage four

The symbol <<< indicates that a rotation to the left of r[i] bits must be executed. The value of r[i] is supplied by the following table, for each stage:

r[0..15] = {7, 12, 17, 22, 7, 12, 17, 22, 7, 12, 17, 22, 7, 12, 17, 22}

r[16..31] = {5, 9, 14, 20, 5, 9, 14, 20, 5, 9, 14, 20, 5, 9, 14, 20}

r[32..47] = {4, 11, 16, 23, 4, 11, 16, 23, 4, 11, 16, 23, 4, 11, 16, 23}

r[48..63] = {6, 10, 15, 21, 6, 10, 15, 21, 6, 10, 15, 21, 6, 10, 15, 21}

The value of k[i] is determined by the following formula:

k[i] = E(abs(sin(i + 1)) × 2^{32}), where E designates the integer part and (i+1) is expressed in radians.

1.2.2. SHA algorithm

The Secure Hash Algorithm (SHA) is associated with multiple hash algorithms: SHA-0, SHA-1 and SHA-2. Algorithm SHA-0 is inspired by the MD5 algorithm. It is not recommended for using for security questions. Algorithm SHA-1 is a modified version of algorithm SHA-0. Algorithm SHA-2 is composed of two algorithms:

– algorithm SHA-256, including versions SHA-224 and SHA-256;

– algorithm SHA-512, including versions SHA-384, SHA-512, SHA-512/256, and SHA-512/224.

1.2.2.1. *SHA-1 algorithm*

The functioning of the SHA-1 algorithm is similar to that of the MD5 algorithm, for which the principal differences introduced are:

– there are 5 registers (A, B, C, D and E) used to produce the 160-bit digest;

– the number of primary iterations increases to 80.

Figure 1.6 shows the primary iteration of the SHA-1 algorithm.

The SHA-1 algorithm uses a Boolean function F among four and changes function every 20 iterations. Likewise, the value of k[i] remains constant for 20 iterations.

The number of words w[g] is equal to 80, one word per iteration. The first 16 words belong to the 512-bit block. The next 64 words are obtained using a rotation applied to the result of an exclusive OR of 4 words calculated during previous iterations.

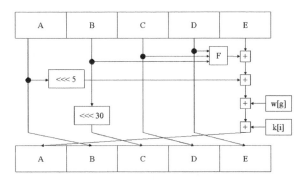

Figure 1.6. *Primary iteration of the SHA-1 algorithm*

1.2.2.2. *SHA-2 algorithm*

Algorithm SHA-226 and its truncated version, SHA-224, are structured on the basis of 32-bit words and a cutting into 512-bit blocks.

Algorithm SHA-512 and its truncated versions, SHA-384, SHA-512/256, and SHA-512/224, use 64-bit words and a cutting into 1024-bit blocks.

The SHA-2 algorithm has the following specific characteristics:

– there are eight registers (A, B, C, D, E, F, G and H). The capacity of each register is 32 bits (as for SHA-1 or MD5) in the case of the SHA-256 algorithm, or 64 bits in the case of the SHA-512 algorithm;

– there are 64 primary iterations in the case of SHA-256 (as for MD5) and 80 in the case of SHA-512 (as for SHA-1);

– the length of the data block is coded at the end of filling on 64 bits in the case of SHA-256 (as for SHA-1 or MD5) and on 128 bits in the case of SHA-512.

The digest of each version is indicated by the last suffix of the algorithm concerned:

– the SHA-256 algorithm produces a 256-bit digest generated by the content of eight registers (A, B, C, D, E, F, G and H);

– the SHA-224 algorithm uses the SHA-256 algorithm and produces a 224-bit digest generated by the content of seven registers (A, B, C, D, E, F and G);

– the SHA-512 algorithm produces a 512-bit digest generated by the content of eight registers (A, B, C, D, E, F, G and H);

– the SHA-384 algorithm uses the SHA-512 algorithm and produces a 384-bit digest generated by the content of six registers (A, B, C, D, E and F);

– the SHA-512/256 algorithm uses the SHA-512 algorithm and produces a 256-bit digest generated by the content of four registers (A, B, C and D;

– the SHA-512/224 algorithm uses the SHA-512 algorithm and produces a 224-bit digest generated by the content of three registers (A, B and C) and the first 32 bits of register D.

Figure 1.7 shows the primary iteration of the SHA-2 algorithm.

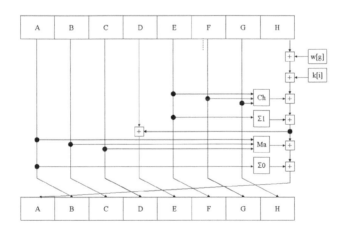

Figure 1.7. *Primary iteration of the SHA-2 algorithm*

The operating principle pertaining to each iteration is identical for all of the algorithms in the SHA-2 family. However, these algorithms differ in the following ways:

– the SHA-256 and SHA-224 algorithms use a specific initialization vector;

– the SHA-512, SHA-384 and SHA-512/256 algorithms use a specific initialization vector;

– the SHA-256 and SHA-224 algorithms, on the one hand, and the SHA-512, SHA-384 and SHA-512/256 algorithms, on the other hand, use Boolean functions (Ch, Ma), rotation operations ($\Sigma 0$, $\Sigma 1$), and specific k[i] constants. Likewise, the calculation of the word w[g] is specific for $g \geq 16$.

1.2.3. *HMAC mechanism*

The keyed-Hash Message Authentication Code (HMAC) mechanism defines the association of secret key K and a data block D, used for the calculation of the MAC seal, using a hash function. The digest obtained, which acts as an MAC seal, enables the checking of the authenticity (origin and integrity) of data block D.

The HMAC mechanism is also used to calculate derived keys.

The MAC seal is calculated using the formula below:

MAC(D) = HMAC(K, D) = H((K0 $\oplus$ opad) | | H((K0 $\oplus$ ipad) | | D))

H is the hash function (MD5 or SHA).

K0 is the key calculated from secret key K.

Outer padding (opad) is the filling composed of byte 5c (in hexadecimals) repeated B times.

Inner padding (ipad) is the filling composed of bite 36 (in hexadecimals) repeated B times.

B is the size of the data block to which the hash function is applied. If the block size of the hash function is 512 bits, ipad and opad are 64 repetitions of bytes 5c and 36 (in hexadecimals).

The sign | | designates the concatenation of bytes.

The calculation of the MAC seal takes place in the following stages:

– Stage 1: the length of key K is equal to B.

K0 = K

Go to stage 4

– Stage 2: the length of key K is greater than B.

Apply the hash function H to key K to obtain a digest of length L.

Complete the digest calculated with zeros to obtain size B.

K0 = H(K) | | 00....00

Go to stage 4

– Stage 3: the length of key K is smaller than B.

Complete key K with zeros to obtain size B.

– Stage 4: execute an exclusive OR between key K0 and ipad to obtain the structure K0 ⊕ ipad.

– Stage 5: concatenate data to authenticate D with the structure generated in stage 4 to obtain the structure K0 ⊕ ipad) | | D.

– Stage 6: apply hash function H to the structure generated in stage 5 to obtain the structure H((K0 ⊕ ipad) | | D).

– Stage 7: execute an exclusive OR between key K0 and opad to obtain the structure K0 ⊕ opad.

– Stage 8: concatenate the structures generated in stages 6 and 7 to obtain the structure (K0 ⊕ opad) || H((K0 ⊕ ipad) || D).

– Stage 9: apply hash function H to the structure generated in stage 8 to obtain the MAC seal enabling authentication of data D.

It is possible to truncate the structure generated in stage 9 to retain a reduced number of left t bits as an MAC seal. For security questions, the number t must be greater than or equal to max(L/2, 80), with L being the length of the digest. The MAC seal is designated in the form of HMAC-H-t. For example, notation HMAC-SHA1-80 designates the calculation of an MAC seal using hash function SHA-1, the digest of which is truncated to 80 bits.

1.3. Key exchange

1.3.1. *Secret-key generation*

The two methods of establishing a secret key are key transport and generation. Key transport generally uses a public key to encrypt the secret key.

The Diffie–Hellman mechanism is used to generate a secret key based on modular exponentiation. The procedure occurs as follows:

– Alice and Bob choose a prime number p (*prime*) with a primitive root g (*generator*) modulo p;

– Alice chooses a secret number a such that $1 \leq a \leq p$-1;

– Bob chooses a secret number b. such that $1 \leq b \leq p$-1;

– Alice calculates the value $A = g^a$ [*mod p*] and sends it to Bob;

– Bob calculates the value $B = g^b$ [*mod p*] and sends it to Alice;

– Alice generates the secret key equal to: $(B \ [mod \ p])^a = B^a \ [mod \ p] = (g^b)^a \ [mod \ p] = g^{ba} \ [mod \ p]$;

– Bob generates the same secret key equal to: $(A \ [mod \ p])^b = A^b \ [mod \ p] = (g^a)^b \ [mod \ p] = g^{ab} \ [mod \ p]$.

A third party can recover the values of p, g, $A = g^a \ [mod \ p]$ and $B = g^b \ [mod \ p]$, which circulate unscrambled in the network. This does not allow the third party to calculate g^{ab} $[mod \ p]$ if the numbers (p, a, b) are large enough to avoid an attack through exhaustive search.

Table 1.3 supplies the formula used to calculate the value of the prime number p (*prime*), with the value of g (*generator*) being fixed at 2, for MODP (MODular exponential modulus P) groups.

Group no.	Designation	Value of p
1	768-bit MODP	$p = 2^{768} - 2^{704} - 1 + 2^{64} \times \{ \ E[2^{638} \times pi] + 149686\}$
2	1024-bit MODP	$p = 2^{1024} - 2^{960} - 1 + 2^{64} \times \{ \ E[2^{894} \times pi] + 129093 \}$
5	1536-bit MODP	$p = 2^{1536} - 2^{1472} - 1 + 2^{64} \times \{ \ E[2^{1406} \times pi] + 741804 \}$
14	2048-bit MODP	$p = 2^{2048} - 2^{1984} - 1 + 2^{64} \times \{ \ E[2^{1918} \times pi] + 741804 \}$
15	3072-bit MODP	$p = 2^{3072} - 2^{3008} - 1 + 2^{64} \times \{ \ E[2^{2942} \times pi] + 169031 \}$
16	4096-bit MODP	$p = 2^{4096} - 2^{4032} - 1 + 2^{64} \times \{ \ E[2^{3966} \times pi] + 240904 \}$
17	6144-bit MODP	$p = 2^{6144} - 2^{6080} - 1 + 2^{64} \times \{ \ E[2^{6014} \times pi] + 929484 \}$
18	8192bit MODP	$p = 2^{8192} - 2^{8128} - 1 + 2^{64} \times \{ \ E[2^{8062} \times pi] + 4743158 \}$

Table 1.3. *MODP Diffie–Hellmann groups*

A second secret-key generation technique is based on the elliptic curve. The procedure occurs as follows:

– Alice and Bob agree on the two public values a and b of an elliptic curve $y^2 = x^3 + ax + b$;

– Alice and Bob agree on a public point P located on the elliptic curve;

– Alice chooses a point on the curve d_a considered to be Alice's secret;

– Alice sends Bob the point constructed using $(d_aP) = d_a + P$;

– Bob chooses a point on the curve d_b considered to be Bob's secret;

– Bob sends Alice the point constructed using $(d_bP) = d_b + P$;

– Alice calculates point $d_a(d_bP) = d_a + d_b + P$, which constitutes the secret key;

– Bob calculates point $d_b(d_aP) = d_b + d_a + P$, which constitutes the same secret key.

1.3.2. *Public key distribution*

Asymmetrical encryption algorithms are based on the diffusion of the public key by the holder of the private key. Conversely, nothing guarantees the identity of the holder of the private key. The certificate, signed by a trusted authority, enables a public key to be associated with a party. All the holders of the public key possess the trusted authority's public key, which enables them to ensure the certificate's authenticity.

Certificates are managed by a public key infrastructure (PKI), which ensures the following functions:

– the user registration;

– the generation of certificates;

– the renewal of certificates;

– the revocation of certificates;

– the publication of certificates;

– the publication of revocation lists.

PKI is divided into several entities:

– the certification authority, whose mission is to sign Certificate Signing Requests (CSR) and Certificate Revocation Lists (CRL);

– the registration authority, whose mission is to generate certificates and to conduct usage verifications of the identity of the final user;

– the repository authority, whose mission is to store digital certificates and revocation lists;

– the key escrow authority, which provides authorities with the means of decrypting data for a user.

A certificate is composed of two parts:

– the data being certified, which contains the base fields and extensions;

– the certificate signature.

The base fields of a certificate provide the following information:

– the structure version of the certificate (value = 3);

– the series number;

– the information about the certificate signature (algorithms and parameters);

– the name of the certificate issuer;

– the validity period of the certificate;

– the name of the certificate holder;

– the public key (value of the public key, algorithm and parameters).

Certificate extensions enable a more precise specification of the following characteristics:

– information about keys;

– information on certification policies;

– additional information on the issuer and holder of the certificate;

– limitations on the certification process.

2

802.1x Mechanism

2.1. General introduction

The 802.1x access control mechanism is deployed in the Local Area Network (LAN) implementing the following technologies:

– Ethernet technology in the case of access to a switch;

– Wireless-Fidelity (Wi-Fi) in the case of a connection to an access point.

There are three components to the 802.1x mechanism (Figure 2.1):

– the supplicant is the device (network host) wishing to access the Ethernet or Wi-Fi network;

– the authenticator is the device (Ethernet switch or Wi-Fi access point) that controls the supplicant's access to the LAN network;

– the authentication server is the device that authenticates the supplicant and authorizes access to the LAN network.

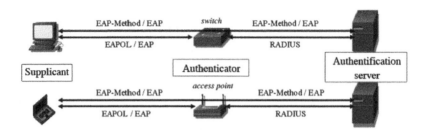

Figure 2.1. *Components of the 802.1x mechanism*

802.1x mechanism relies on a set of protocols (Figure 2.2):

– the Extensible Authentication Protocol (EAP) Over LAN (EAPOL), exchanged between the supplicant and the authenticator;

– the EAP protocols exchanged between the supplicant on one hand and the authenticator or authentication server on the other. The EAP protocol is carried by the EAPOL protocol on the interface between the supplicant and the authenticator. The EAP protocol carries EAP-*Method* messages;

– the Remote Authentication Dial-In User Service (RADIUS) protocol, exchanged between the authenticator and the authentication server. The RADIUS protocol carries the EAP protocol on the interface between the authenticator and the authentication server.

2.2. EAPOL protocol

The EAPOL protocol is exchanged between the supplicant and the authenticator. It initiates the supplicant's identity announcement and the capacities of each end. It ensures the transport of EAP/EAP-*Method* messages, which enable authentication of the supplicant, and possibly of the authentication server.

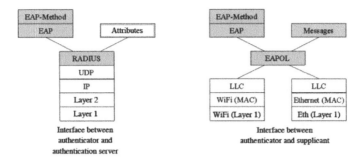

Figure 2.2. *Protocol architecture*

The structure of the EAPOL protocol is shown in Figure 2.3.

The EAPOL protocol is made up of a 4 byte header and a packet body.

Version: this field, coded on one byte, identifies the version of the protocol and has the value of 03 in hexadecimals for the latest standardized version.

Packet type: this field, coded on one byte, identifies the type of data encapsulated by the EAPOL header. The EAPOL header can encapsulate an EAPOL message or an EAP message.

Packet Body Length: this field, coded on two bytes, indicates the size of the data encapsulated by the EAPOL header.

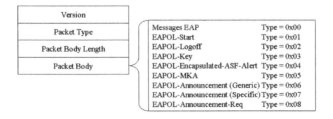

Figure 2.3. *Structure of EAPOL message*

2.2.1. *EAPOL-Start message*

The EAPOL-*Start* message was transmitted without a message body in version 2 of the protocol. In version 3, the EAPOL-*Start* message can be transmitted with or without a message body. This message, transmitted by the supplicant, is used to initialize the 802.1x mechanism.

If the least significant bit of the first byte of the message body is set at ONE, the receiver of the EAPOL-*Start* message must make an announcement. The other bits of the first byte are set at zero.

The other bytes of the message body, if present, have a Type, Length, Value (TLV) structure, which gives information about network access conditions.

2.2.2. *EAPOL-Logoff message*

The EAPOL-*Logoff* message is transmitted without a message body. This message, transmitted by the supplicant, is used to terminate the 802.1x mechanism. At the end of this message, the supplicant is no longer authenticated and its access to the LAN network is blocked.

2.2.3. *EAPOL-Key message*

The EAPOL-*Key* message is transmitted by the supplicant or by the authenticator. It is used for the establishment of authentication and encryption keys derived from a master key. The EAPOL-*Key* message is transmitted with a *Key Descriptor* message body containing the information necessary for key establishment. This message body is described in Chapter 3 for the Wi-Fi Protected Access (WPA) mechanism.

2.2.4. *EAPOL-Encapsulated-ASF-Alert message*

The   EAPOL-*Encapsulated-ASF-Alert*   message   is transmitted by the supplicant during authentication. It usually contains information specific to each constructor.

2.2.5. *EAPOL-MKA message*

The MACsec Key Agreement (EAPOL-MKA) message was introduced in version 3 for negotiating the association of Ethernet security. This point is not addressed in this book.

2.2.6. *EAPOL-Announcement message*

The EAPOL-*Announcement* message was introduced in version 3 for the transmission by the authenticator of information concerning network access conditions. The message body is composed of TLV structures.

The *Network Identity* TLV structure contains the identification of the network being subjected to access control. This structure can also be present in EAPOL-*Start* and EAPOL-*Announcement-Req* messages.

The *Access Information* TLV structure contains access information (access status, EAPOL messages processed). This structure can also be present in EAPOL-*Start* and EAPOL-*Announcement-Req* messages.

The *MACsec Cipher Suites* TLV structure contains information on the encryption algorithms supported by the authenticator for the MACsec mechanism.

The *Key Management Domain* TLV structure contains the key-management domain name associated with a network name.

2.2.7. *EAPOL-Announcement-Req message*

The EAPOL-*Announcement-Req* message was introduced in version 3. If the message body is absent or the least significant bit of the first byte of the message body is positioned at one, the receiver of the EAPOL-*Announcement-Req* message must make an announcement. As for the EAPOL-*Start* message, the other bytes of the message body, if present, have a TLV structure that gives information on access conditions. Unlike the EAPOL-*Start* message, the EAPOL-*Announcement-Req* message does not initiate the authentication procedure.

2.3. EAP protocol

The EAP protocol is deployed for the access of a supplicant (network host) to the authenticator (switch or access point) and the authentication server. It enables the transport of authentication data and does not require Internet Protocol (IP) connectivity.

The structure of the EAP protocol is shown in Figure 2.4.

The EAP protocol is composed of a 4 byte header and possibly an EAP-*Method* message.

Code: this field, coded on one byte, identifies the type of EAP message:

– *Request*: this message enables the authenticator to communicate with the supplicant, for example, to transmit to it an EAP-*Method* message received from the authentication server. The supplicant can also transmit this message to the authenticator to request its identity during mutual authentication.

– *Response*: this message is sent in reply to the previous message. It can correspond, for example, to the supplicant's authentication data contained in an EAP-*Method* message.

– *Success*: this message is used by the authenticator to inform the supplicant that it has been successfully authenticated;

– *Failure*: this message is used by the authenticator to inform the supplicant that authentication has failed.

Identifier: this field, coded on one byte, enables correlation of the messages exchanged between the supplicant and the authenticator.

Length: this field, coded on two bytes, indicates the size of the EAP message. This value is identical to the value of the *Packet Body Length* field in the EAPOL protocol header.

When the EAP message is a request or a reply, the header encapsulates an EAP-*Method* message that contains the *Type* field, coded on one byte, identifying the type of data in the EAP-*Method* message. The first three values in the *Type* field are reserved for specific messages (*Identity*, *Notification* and NAK). The other values pertain to identification methods such as, EAP- message digest 5 (MD5) (*Type* = 4), EAP- Transport Layer Security (TLS) (*Type* = 13) or EAP-Tunneled TLS (TTLS) (*Type* = 21).

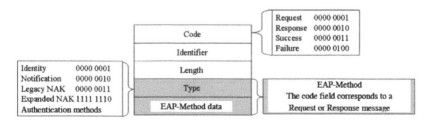

Figure 2.4. *EAP message structure*

For the two other EAP messages (*Success* or *Failure*), only the four-byte header is transmitted, without an associated EAP-*Method* message.

Table 2.1 shows the value of the *Type* field for the different methods used.

Type	EAP-Method
1	*Identify*
2	*Notification*
3	NAK (response only)
4	MD5-Challenge
5	One Time Password (OTP)
6	Generic Token Card (GTC)
7	Non-attributed
8	Non-attributed
9	RSA Public Key Authentication
10	DSS Unilateral
11	KEA
12	KEA-VALIDATE
13	EAP-TLS
14	Defender Token (AXENT)
15	RSA Security SecurID EAP
16	Arcot Systems EAP
17	EAP-Cisco Wireless (LEAP)
18	Nokia IP Smart Card Authentication
19	SRP-SHA1 Part 1
20	SHP-SHA2 Part 2
21	EAP-TTLS
22	Remote Access Service
23	UTMS Authentication and Key Agreement
24	EAP-3Com Wireless
25	PEAP
26	MS-EAP Authentication
27	Mutual Authentication w/Key Exchange (MAKE)
28	CRYPTOCard
29	EAP-MSCHAP-V2
30	DynamID
31	Rob EAP
32	SecurID EAP
33	EAP-TLV
34	SentriNET
35	EAP-Actiontec Wireless

Type	EAP-Method
36	Cogent Systems Biometrics Authentication EAP
37	AirFortress EAP
38	EAP-HTTP Digest
39	SecureSuite EAP
40	Device Connect EAP
41	EAP-SPEKE
42	EAP-MOBAC
43	EAP-FAST
44–191	Non-attributed; possibility of attribution by the IANA organization
192–253	Reserved
254	Extension of *Type* field
255	Experimental use

Table 2.1. *Different types of EAP-Method messages*

2.3.1. *EAP-Method Identity*

The EAP-*Method Identity* message is used before or during the authentication phase. It is used to transport the supplicant's identity and, in some cases, during mutual authentication, the authenticator's identity. It can include data that will be presented to the user.

The EAP-*Method Identity* message is carried by the EAP *Request* message for the identity request; then by the EAP *Response* message for the reply containing the identity.

The EAP-*Method Identity* message is transferred to the authentication server. If the identity received is invalid, this operation can be repeated several times. It is also possible for the authenticator to verify the supplicant's identity.

2.3.2. *EAP-Method Notification*

The EAP-*Method Notification* message is used before or during the authentication phase. It is used to transport information on authentication status. It includes data that will be presented to the user.

The EAP-*Method Notification* message is carried by the EAP *Request* message for the notification request; then by the EAP *Response* message for the notification response.

2.3.3. *EAP-Method NAK*

The EAP-*Method Legacy* NAK message is used by the supplicant to indicate that it does not support the authentication method suggested by the authentication server.

The EAP-*Method Legacy* NAK message is carried by the EAP *Response* message. It contains the authentication methods supported by the supplicant. If the message contains a *Type* field equal to zero. This means that the supplicant rejects the suggestion and that there is no alternative.

The EAP-*Method Expanded* NAK message is used by the supplicant in response to a request that also contains the EAP-*Method Expanded* NAK message. This functionality enables the number of types of authentication methods to be expanded beyond the 255 values allowed by the *Type* field.

2.4. RADIUS protocol

The RADIUS protocol is used for transporting EAP-*Method* messages between the authentication server and the authenticator (switch or access point), used to identify the supplicant. RADIUS messages are encrypted and checked in their entirety using a secret shared between the two end points.

The structure of the RADIUS protocol is shown in Figure 2.5.

The RADIUS message contains a 20-byte header and a set of attributes.

Code: this field, coded on one byte, identifies the type of RADIUS message:

– Code = 1: RADIUS Access-Request;

– Code = 2: RADIUS Access-Accept;

– Code = 3: RADIUS Access-Reject;

– Code = 11: RADIUS Access-Challenge

Identifier: this field, coded on one byte, is used to correlate RADIUS *Access-Request* and *Access-Challenge* messages. If the authentication server has not responded to the RADIUS *Access-Request* message, the authenticator transmits this message again with the same value for this field.

Length: this field, coded on two bytes, indicates the size of the RADIUS message.

Authenticator: this field is coded on 16 bytes. For the RADIUS *Access-Request* message transmitted by the authenticator, this field, *Request Authenticator,* contains a random number. This value changes for each new value of the *Identifier* field.

For the response from the authentication server (RADIUS *Access-Accept*, RADIUS *Access-Reject* and RADIUS *Access-Challenge*), this field, *Response-Authenticator*, is calculated in the following manner:

– the secret shared between the authenticator and the authentication server is associated with the RADIUS *Access-Request* message;

– the 16 byte digest is calculated from the structure obtained, using the MD5 hash function.

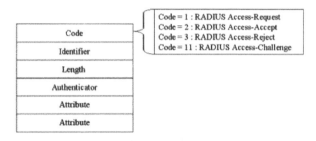

Figure 2.5. *Structure of RADIUS message*

2.4.1. *RADIUS messages*

2.4.1.1. *Access-Request message*

The RADIUS *Access-Request* message is transmitted by the authenticator to the authentication server. This message is used during supplicant authentication. It generally contains the *User-Name*, *User-Password*, and Narrow-band Access Server (*NAS*)-*IP-Address* attributes.

2.4.1.2. *Access-Challenge message*

The RADIUS *Access-Challenge* message is one of the authentication server's responses to the RADIUS *Access-Request* message. In this message, the authentication server suggests a challenge to the supplicant. This message can contain one or more of the following attributes: *Vendor-Specific*, *Idle-Timeout* and *Session-Timeout*.

2.4.1.3. *Access-Accept message*

The RADIUS *Access-Accept* message is one of the authentication server's responses to the RADIUS *Access-Request* message. In this message, the authentication server accepts the supplicant's authentication request. This message can contain one or more attributes used for the

RADIUS *Access-Challenge* message. Upon receipt of this message, the authenticator unblocks the supplicant's access.

2.4.1.4. *Access-Reject message*

The RADIUS *Access-Reject* message is one of the authentication server's responses to the RADIUS *Access-Request* message. In this message, the authentication server rejects the supplicant's authentication request. This message can contain one or more *Reply-Message* attributes. Upon receipt of this message, the authenticator continues to block the supplicant's access.

2.4.2. *RADIUS attributes*

RADIUS attributes have a TLV structure. The *Type* field, coded on one byte, identifies the attribute. The *Length* field, coded on one byte, provides the size of the attribute. The *Value* field, of variable size, contains various structures of data:

– a text composed of 1–253 bytes of characters in Universal Character Set (UCS) Transformation Format 8 bits (UTF-8) format;

– a string composed of 1–253 bytes of binary data;

– an IP version 4 (IPv4) address coded on 32 bits;

– an integer number coded on 32 bits.

2.4.2.1. *EAP-Message attribute*

The EAP-*Message* attribute, identified by the *Type* field as equal to 79, transports EAP messages exchanged between the supplicant and the authentication server, which enable the authenticator to avoid analyzing the EAP message. These EAP messages are transmitted by the authenticator in a RADIUS *Access-Request* message and the authentication server in RADIUS *Access Challenge*, *Access-Accept* and *Access-Reject* messages.

2.4.2.2. *Message-Authenticator attribute*

The *Message-Authenticator* attribute, identified by the *Type* field as equal to 80, is used to authenticate RADIUS messages and check integrity. The *Value* field contains a 16 byte field calculated using the Hashed message authentication code (HMAC)-MD5 function.

2.4.2.3. *Password-Retry attribute*

The *Password-Retry* attribute, identified by the *Type* field as equal to 75, is included in a RADIUS *Access-Reject* message to indicate the maximum number of authentication attempts the supplicant is permitted to make. This value is coded on 4 bytes and is located in the *Value* field.

2.4.2.4. *User-Name attribute*

The *User-Name* attribute, identified by the *Type* field as equal to 1, contains the name of the authenticator. It is transmitted in the RADIUS *Access-Request* message.

2.4.2.5. *User-Password attribute*

The *User-Password* attribute, identified by the *Type* field as equal to 2, contains the authenticator seal, calculated as follows:

– the secret shared between the authenticator and the authentication server is completed by zeros to constitute a structure whose size is a multiple of 16 bytes;

– the MD5 hash function is applied to the secret associated with the *Request Authenticator* field;

– an exclusive OR is executed between the result and the first 16 bytes of the secret.

This attribute is transmitted in a RADIUS *Access-Request* message. It can be used in combination with the *Message-Authenticator* attribute, which improves resistance to attacks.

2.4.2.6. *NAS-IP-Address attribute*

The NAS-IP-*Address* attribute, identified by the *Type* field as equal to 4, contains the IPv4 address of the authenticator. This attribute is only transmitted in the RADIUS *Access-Request* message.

2.4.2.7. *NAS-Port attribute*

The NAS-*Port* attribute, identified by the *Type* field as equal to 5, contains the physical interface number of the authenticator, or switch, to which the supplicant is connected. This attribute is only transmitted in the RADIUS *Access-Request* message.

2.4.2.8. *Service-Type attribute*

The *Service-Type* attribute, identified by the *Type* field as equal to 6, contains the type of service requested by the supplicant in the RADIUS *Access-Request* message, or the service granted by the authentication server in the RADIUS *Access-Accept* message. If the service requested by the supplicant is not supported, the authentication server responds with a RADIUS *Access-Reject* message.

2.4.2.9. *Vendor-Specific attribute*

The *Vendor-Specific* attribute, identified by the *Type* field as equal to 26, enables a constructor to implement specific attributes. If the authentication server is not able to interpret this attribute, it ignores it. For its part, the authenticator must then adapt and function in degraded mode.

2.4.2.10. *Session Timeout attribute*

The *Session-Timeout* attribute, identified by the *Type* field as equal to 27, sets the duration during which the interface with the supplicant is active. This attribute is transmitted in RADIUS *Access-Accept* or *Access-Challenge* messages.

2.4.2.11. *Idle-Timeout attribute*

The *Idle-Timeout* attribute, identified by the *Type* field as equal to 28, sets the duration during which the interface with the supplicant is in the awake state before final disconnection. This attribute is transmitted in RADIUS *Access-Accept* or *Access-Challenge* messages.

2.4.2.12. *Termination-Action attribute*

The *Termination-Action* attribute, identified by the *Type* field as equal to 29, indicates what action the authenticator must take when the service is terminated. This attribute is only transmitted in the RADIUS *Access-Accept* message.

The *Value* field has all of its bits set at zero or one. If all bits are set at one, the authenticator must send a new RADIUS *Access-Request* message at the end of the service.

2.5. Authentication procedures

Prior to the 802.1x authentication procedure, the supplicant must connect to the authenticator:

– if the supplicant is connected to an Ethernet switch, the 802.1x procedure starts when its interface is activated;

– if the supplicant is connected to a Wi-Fi access point, the 802.1x procedure starts at the end of the association phase with the Wi-Fi access point.

Whatever the authentication method used, the procedure is initiated by the supplicant, which transmits the

EAPOL-*Start* message. The authenticator continues the procedure by sending the EAP-*Request* message containing the EAP-*Method Identity* message. The supplicant provides its identity by replying with an EAP *Response*/EAP-*Method Identity* message. This message is transmitted by the authenticator to the authentication server in a RADIUS *Access-Request* message (Figure 2.6).

The next series of operations depend on the authentication method chosen.

After the authentication phase exchanges, the server transmits to the supplicant:

– the EAP *Success* message if it is authenticated, in which case the authenticator authorizes traffic from the supplicant;

– the EAP *Failure* message in the opposite case, and access to the network remains prohibited.

The EAP *Success* (or EAP *Failure*) message is transmitted in a RADIUS *Access-Accept* (or *Access-Reject*) message at the interface between the authentication server and the authenticator (Figure 2.6).

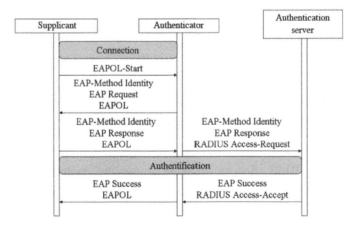

Figure 2.6. *Common exchanges in the authentication procedure*

2.5.1. *EAP-MD5 procedure*

The EAP-MD5 procedure is based on a shared secret. It enables the authentication server to authenticate the supplicant.

The EAP-MD5 procedure is shown in Figure 2.7.

The procedure includes the following operations:

– the authentication server transmits a challenge to the supplicant;

– the supplicant responds with a seal calculated using the secret and the challenge;

– the authentication server compares the seal received with the seal calculated locally. If the two seals match, the supplicant is authenticated.

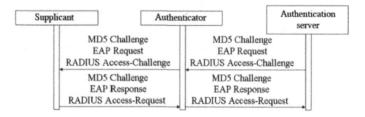

Figure 2.7. *EAP-MD5 authentication procedure*

The RADIUS *Access-Challenge* message generated by the authentication server is encapsulated by an EAP *Request* message.

The RADIUS *Access-Request* message received by the authentication server is encapsulated by an EAP *Response* message.

2.5.2. *EAP-TLS procedure*

The EAP-TLS procedure is based on the TLS protocol described in Chapter 5. The EAP-TLS procedure uses only the handshake phase.

The EAP-TLS procedure provides services of mutual authentication, algorithm negotiation for integrity checking and confidentiality and key exchange between two end points.

The EAP-TLS procedure is described in Figure 2.8.

Upon receipt of the EAP-*Method Identity* message, the initialization phase starts with the TLS_*hello_request* message transmitted by the authentication server.

Thus, communication between the supplicant and the authentication server is engaged and the discovery phase starts. The supplicant responds with a TLS_*client_hello* message.

The TLS_*client_hello* message contains the version number of the TLS protocol, a session identifier, a random number and series of algorithms supported by the supplicant.

The random number is formed by the concatenation of time in seconds since January 1, 1970 (4 bytes) and 28 random bytes.

The next algorithm series contains the key exchange algorithm, the encryption algorithm, the key length and the integrity-checking algorithm.

The authentication server terminates the discovery phase by sending the TLS_*server_hello* message and starts the exchange phase with the following messages:

– TLS_*certificate* containing the certificate enabling certificate authentication;

– TLS_*server_key_exchange* containing encryption information necessary for the creation of the *PreMaster* key;

– TLS_*certificate_request* requesting the supplicant to transmit its certificate;

– TLS_*server_hello_done* terminating the discovery phase. This message contains no information.

The TLS_*server_hello* message contains the version number of the TLS protocol, another random number, a session identifier and a series of algorithms.

The authentication server determines the value of the session identifier if the value indicated by the supplicant has a value of zero or if it is not recognized by the authentication server. In the opposite case, the authentication server uses the value indicated by the supplicant.

Likewise, the authentication server must choose a set of algorithms from among the ones suggested by the supplicant.

The supplicant checks the certificate provided by the authentication server and responds to the previous message by sending the following messages:

– TLS_*certificate* containing the certificate enabling authentication of the supplicant;

– TLS_*client_key_exchange* containing encryption information necessary for the creation of the *PreMaster* key;

– TLS_*certificate_verify* containing a digest of the concatenation of all messages exchanged from the TLS_*client_hello* message to the message preceding the TLS_*certificate_verify* message. The digest is encrypted with the supplicant's private key;

– TLS_*change_cipher_spec* indicating that after this message, all exchanges will use the new *Master* key and the new algorithms that have just been negotiated. The *Master* key is derived from the *PreMaster* key;

– TLS_*finished* containing a digest of the concatenation of all messages exchanged since the TLS_*client_hello* message. This message is encrypted and completely controlled.

The next phase involves the server's indication of the implementation of the key and encryption algorithms. The authentication server transmits the following messages to the supplicant:

– TLS_*change_cipher_spec* indicating that after this message, all exchanges will use the new *Master* key and the new algorithms that have just been negotiated;

– TLS_*finished* containing a digest of the concatenation of all messages exchanged since the TLS_*client_hello* message. This message is encrypted and completely controlled.

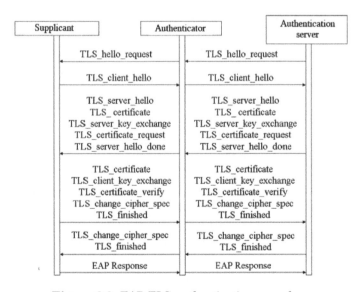

Figure 2.8. *EAP-TLS authentication procedure*

2.5.3. *EAP-TTLS procedure*

The EAP-TTLS procedure is an evolution of the EAP-TLS procedure. It enables mutual authentication by the supplicant and the authentication server, but it does not require the supplicant to possess a certificate.

The EAP-TTLS procedure uses the two phases of the TLS protocol; the handshake phase and the data transfer phase.

During the handshake phase, the authentication server authenticates itself to the supplicant using the TLS procedure. Keys are generated in order to establish a secure tunnel between the supplicant and the authenticator, for exchanges of information pertaining to the data transfer phase.

During the data transfer phase, the supplicant is authenticated by the authentication server. The authentication protocol using the tunnel is negotiated.

The supplicant's identity is provided only during the second phase, which ensures its confidentiality. Before the implementation of the EAP-TTLS procedure, the authentication phase must normally start with the presentation of the supplicant's identity. In this case, the initial EAP-*Method Identity* message contains no information.

The first phase of the EAP-TTLS procedure is shown in Figure 2.9.

The authentication server initializes the EAP-TTLS procedure with the TTLS_*hello_request* message. This indicates that the supplicant must start the handshake phase. The procedure is similar to the EAP-TLS procedure. Note that the client does not transmit a certificate to the authentication server.

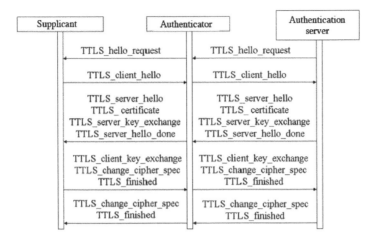

Figure 2.9. *EAP-TTLS authentication procedure:*
server authentication phase

The second phase of the EAP-TTLS procedure is shown in Figure 2.10.

The secure tunnel established between the supplicant and the authenticator enables the transfer of authentication data from the supplicant, encapsulated in Attribute-Value Pairs (AVP) sequences. The supplicant authentication method (for example, the EAP-MD5 method) is defined at this level.

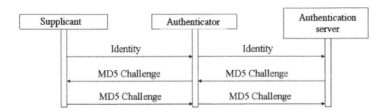

Figure 2.10. *EAP-TTLS authentication procedure:*
supplicant authentication phase

WPA Mechanisms

3.1. Introduction to Wi-Fi technology

A Wireless-Fidelity (Wi-Fi) network is a collection of Basic Service Set (BSS) cells. A BSS cell is the radio coverage zone of an Access Point (AP). It is identified by a BSS identifier (BSSID), which is usually the Medium Access Control (MAC) address of the AP (Figure 3.1).

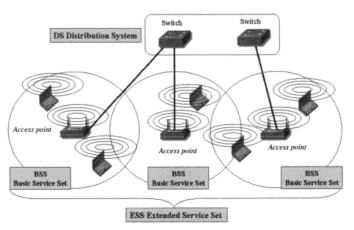

Figure 3.1. *Wi-Fi network architecture*

A whole set of APs constitutes an Extended Service Set (ESS) network, identified by a Service Set Identifier (SSID).

The APs are connected to switches in the Local Area Network (LAN), called a Distribution System (DS), by an Ethernet interface (Figure 3.1).

Wi-Fi technology defines two layers (Figure 3.2): the physical layer and the MAC data link layer. The MAC data link layer is completed by another Logical Link Control (LLC) data link layer, the main contribution of which is the insertion of the *Type* field, which defines the type of data encapsulated.

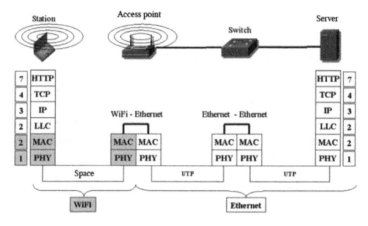

Figure 3.2. *Protocol architecture of Wi-Fi network*

The AP ensures the modification of the physical layer and MAC data link layer between the Wi-Fi interface on the one hand and Ethernet interface on the other hand. It maintains a switching table where the registered stations are listed.

Station registration by the AP occurs in three phases: scanning, authentication and association.

Scanning can be passive or active. When it is passive, the station scans the BEACON channel on each radio channel containing the characteristics of the cell.

When scanning is active, the station generates the PROBE_REQUEST management message broadcast to all APs using a broadcast BSSID. The AP responds to the request with the PROBE_RESPONSE management message containing the characteristics of the cell.

Two modes are defined for the authentication phase:

– Open System Authentication (OSA) mode, used for Wi-Fi Protected Access (WPA1) and WPA2 mechanisms;

– Shared Key Authentication (SKA) mode, used only for the Wired Equivalent Privacy (WEP) mechanism.

In OSA mode, authentication takes place in two stages:

– the station sends the AUTHENTICATION management message mentioning the authentication mode in it;

– the AP responds with the AUTHENTICATION management message containing the authentication status (success or failure).

In SKA mode, authentication takes place in four stages:

– the station sends the AUTHENTICATION management message mentioning the authentication mode in it;

– the AP sends the AUTHENTICATION management message containing a string of characters;

– the station generates the AUTHENTICATION management message containing the encrypted character string;

– the AP verifies the station response and sends the AUTHENTICATION management message containing the authentication status (success or failure).

The purpose of the association phase is to check whether the transmission characteristics of each party (station and AP) are compatible or not. It takes place in two phases:

– the station sends the ASSOCIATION_REQUEST management message containing the characteristics it supports;

– the AP sends the ASSOCIATION_RESPONSE management message containing the characteristics retained and the association status (success or failure).

Cell change takes place at the initiative of the station, via the emission of the REASSOCIATION_REQUEST management message toward a new AP. The new AP responds with the REASSOCIATION_RESPONSE management message. The station must authenticate itself to the new AP.

3.2. Security mechanisms

Radio interface security started with the WEP mechanism. Due to its weaknesses, it was supplanted by the WPA1 mechanism, and then by the WPA2 mechanism. WPA1 and WPA2 mechanisms constitute Robust Security Network (RSN) architecture.

The three mechanisms WEP, WPA1 and WPA2 specifically implement third-party access control and data protection (confidentiality and integrity checking) services.

For the WEP mechanism, third-party access control is based on the Rivest Cipher 4 (RC4) algorithm. Access control takes place during the authentication phase, which is a procedure associated with the MAC data link protocol.

WPA1 and WPA2 mechanisms use the 802.1x mechanism described in Chapter 2 for access control. The authentication phase is preceded by the procedure putting the security policy in agreement between the AP and the station.

For WEP and WPA1 mechanisms, encryption is executed by the RC4 algorithm. For the WEP mechanism, the Master Key (MK) is used for the encryption of each Wi-Fi frame. For the WPA1 mechanism, encryption is obtained using a temporary key derived from the MK. In association with encryption, a protocol is added to the MAC data link layer:

– the WEP protocol in the case of the WEP mechanism;

– the Temporal Key Integrity Protocol (TKIP) in the case of the WPA1 mechanism.

For the WPA2 mechanism, encryption is based on the Advanced Encryption Standard (AES) algorithm, and the header of the MAC data link protocol is completed by the Counter-mode/Cipher block Chaining MAC (Message Authentication Code) Protocol (CCMP) header.

Integrity control is provided by a Cyclic Redundancy Check (CRC) for the WEP mechanism.

In the case of the WPA1 mechanism, integrity control uses the MICHAEL algorithm.

In the case of the WPA2 mechanism, integrity control is obtained using the AES encryption algorithm.

3.3. Security policies

The security policies supported by the AP are transmitted in BEACON and PROBE_RESPONSE messages during the listening phase.

The station's response to the security policies supported is included in the ASSOCIATION_REQUEST message during the association phase. This message is validated by the ASSOCIATION_RESPONSE message from the AP.

Security policy information is sent in the RSN Information Element (IE) (Figure 3.3).

Element ID	Length	Version	Group Cipher Suite	Pairwise Cipher Suite Count	Pairwise Cipher Suite List	AKM Suite Count	AKM Suite List	RSN Cap.	PMKID Count	PMKID List

Figure 3.3. *Format of the RSN information element*

The *Element ID* field is coded on 1 byte. It contains the RSN IE identifier. It has a value of 48 in hexadecimals.

The *Length* field is coded on 1 byte. It indicates the size in bytes of all of the fields following the *Length* field.

The *Version* field is coded on two bytes. It indicates the version of the RSN IE. The value of this field is equal to 1.

The *Group Cipher Suite* field is coded on 4 bytes. It contains the selector of the mechanism used for the protection of broadcast and multicast flows.

The *Pairwise Cipher Suite Count* field is coded on 2 bytes. It indicates the number of selectors contained in the *Pairwise Cipher Suite List* field.

The *Pairwise Cipher Suite List* field is coded on 4 bytes per selector of the mechanism used for the protection of unicast flows.

The mechanism selector contains the following fields (Table 3.1):

– the organizationally unique identifier (OUI) field, coded on 3 bytes. It has a value of 00-0F-AC in hexadecimals;

– the *Type* field, coded on 1 byte.

OUI field	*Type* field	Designation
00-0F-AC	0	Group Cipher Suite
00-0F-AC	1	WEP (40-bit key)
00-0F-AC	2	TKIP
00-0F-AC	3	Reserved
00-0F-AC	4	CCMP
00-0F-AC	5	WEP (104-bit key)
00-0F-AC	6–255	Reserved

Table 3.1. *Selector format*

The *AKM Suite Count* field is coded on 2 bytes. It indicates the number of selectors contained in the *AKM Suite List* field. The term Authentication and Key Management (AKM) designates the key management mechanism and the type of authentication mechanism.

The *AKM Suite List* field is coded on 4 bytes per selector pertaining to key management and the type of authentication mechanism (Table 3.2).

OUI field	*Type* field	Designation	
		Authentication	Key management
00-0F-AC	0	Reserved	Reserved
00-0F-AC	1	802.1x	See section 3.4
00-0F-AC	2	PSK	See section 3.4
00-0F-AC	3–255	Reserved	Reserved

PSK: preshared key replacing the Master Key (MK) generated by the authentication server.

Table 3.2. *AKM Suite selector format*

The *RSN Capabilities* field is coded on 2 bytes. It contains the following fields (Figure 3.4):

– *Preauthentication*: the AP positions this bit at one if it supports preauthentication, and at zero in the opposite case. The station always positions this bit at zero. Preauthentication is used when the station must authenticate itself to multiple APs during a handover;

– *No Pairwise*: the station positions this bit at zero if it supports WEP and WPA mechanism keys, and at one in the opposite case. The AP always positions this point at zero;

– *PTKSA Replay Counter*: the value of these two bits determines the number of replay counters associated with unicast traffic. The term Pairwise Transient Key Security Association (PTKSA) designates the context established at the end of the four-way handshake procedure described in section 3.4.3. The term Pairwise Transient Key (PTK) designates the key derived from a Pairwise Master Key (PMK). The hierarchy of keys is described in section 3.4.1.

– *GTKSA Replay Counter*: the value of these two bits determines the number of replay counters associated with broadcast or multicast traffic. The term Group Transient Key Security Association (GTKSA) designates the context established at the end of the group key handshake procedure described in section 3.4.4. The term Group Transient Key (GTK) designates the key derived from the Group Master Key (GMK).

bit	0	1	2 to 3	4 to 5	6 to 15
	Pre Auth.	No Pairwise	PTKSA Replay Counter	GTKSA Replay Counter	Reserved

Figure 3.4. *RSN capabilities field format*

The Pairwise Master Key Identifier *(PMKID) Count* field is coded on 2 bytes. It indicates the number of PMK key PMKID identifiers contained in the *PMKID List* field.

The PMKID identifier is coded on 16 bytes. It is defined using the PMK key, the name of the PMK key, the Authenticator Address (AA) and the Supplicant Address (SPA) in the following manner:

PMKID = HMAC-SHA1-128 (PMK, "PMK Name" || AA || SPA)

The *PMKID Count* and *PMKID List* fields are included by the station in the REASSOCIATION_REQUEST message sent to the new AP during a cell change.

3.4. Key management

3.4.1. *Key hierarchy*

Data protection is based mainly on secret keys. When a security association is established after successful authentication, temporary keys are created from the MK. These derived keys are regularly updated until the context is closed.

The derivation of the PMK key uses the Hashed Message Authentication Code - Secure Hash Algorithm (HMAC-SHA1) function, the result of which is 384 bits in size for the CCMP mechanism and 512 bits in size for the TKIP mechanism.

The PTK key, derived from the PMK key, is obtained using the MAC addresses of the AA and SPA and random numbers (ANonce and SNonce) exchanged during the four-way handshake procedure.

PTK = HMAC-SHA1 (PMK, "Pairwise key expansion", Min (AA, SPA) || Max (AA, SPA) || Min (ANonce, SNonce) || Max (ANonce, SNonce))

The PTK key is cut up in order to provide the following keys:

– 128-bit Key Confirmation Key (KCK). This key is used to authenticate messages during the four-way handshake procedure;

– 128-bit Key Encryption Key (KEK). This key is used to encrypt messages during the four-way handshake and group key handshake procedures;

– 128-bit Temporary Key (TK). This key, used for TKIP and CCMP, is used to encrypt unicast data;

– 64-bit Temporary Message Integrity Code (MIC) Keys (TMK1) and TMK2 keys. These keys, used for TKIP, check the integrity of unicast data. Each direction of transmission uses a specific key in order to generate the seal: TMK1 is used by the AP, and TMK2 is used by the station.

The derivation of the GMK key also uses the HMAC-SHA1 function, with a result 128 bits in size for the CCMP mechanism and 256 bits in size for the TKIP mechanism.

The GTK key, derived from the GMK master key, is obtained from the authenticator's MAC address (AA), and from the random number (Gnonce) obtained during the group key handshake procedure.

GTK = HMAC-SHA1 (GMK, "Group key expansion" || AA || GNonce)

The GTK key is cut up in order to provide the following keys:

– 128-bit Group Encryption Key (GEK). This key, used for TKIP and CCMP, encrypts broadcast and multicast data;

– The 128-bit Group Integrity Key (GIK). This key, used for TKIP, checks the integrity of broadcast and multicast data.

3.4.2. EAPOL-key messages

Extensible Authentication Protocol Over LAN (EAPOL)-*Key* messages are exchanged between the AP and station in order to establish security associations enabling the installation of derived keys.

EAPOL-*Key* messages are used for four-way handshake and group key handshake procedures.

The format of the EAPOL-*Key* message is shown in Figure 3.5.

Descriptor Type – 1 byte	
Key Information – 2 bytes	Key Length – 2 bytes
Key Replay Counter – 8 bytes	
Key Nonce – 32 bytes	
EAPOL-Key IV – 16 bytes	
Key RSC 8 bytes	
reserved – 8 bytes	
Key MIC – 16 bytes	
Key Data Length – 2 bytes	Key Data – n bytes

Figure 3.5. *EAPOL-Key message format*

The *Descriptor Type* field defines the type of EAPOL message. Its value is equal to 254.

The *Key Information* field contains the following indications (Figure 3.6):

– *Key Descriptor Version*. These three bits define the descriptor type of the algorithms used for EAPOL-*Key* messages:

  - a value of 1 indicates the use of HMAC-MD5 for integrity checking and RC4 for encryption,

  - a value of 2 indicates the use of HMAC-SHA1-128 for integrity checking and AES for encryption;

– *Key Type*. This bit, positioned by the AP, indicates whether the message is used for the group key handshake procedure (bit at zero) or the four-way handshake procedure (bit at one);

– *Install*. This bit indicates whether the PTK key can be used (bit at one) or not (bit at zero);

– *Key Ack*. This bit, positioned by the AP, indicates whether a response must be given to this message (bit at one) or not (bit at zero);

– *Key MIC*. This bit indicates the presence of a seal (bit at one) or not (bit at zero). The term MIC designates the seal used for integrity checking;

– *Secure*. This bit indicates whether PTK and GTK keys have been installed (bit at one) or not (bit at zero);

– *Error*. This bit is positioned at one by the station to indicate that an MIC error has been detected in the TKIP data;

– *Request*. This bit is positioned at one by the station to restart either the four-way handshake procedure or the group key handshake procedure;

– *Encrypted Key Data*. This bit indicates whether the *Key Data* field of the EAPOL-*Key* message is encrypted (bit at one) or not (bit at zero).

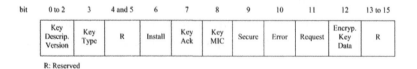

bit	0 to 2	3	4 and 5	6	7	8	9	10	11	12	13 to 15
	Key Descrip. Version	Key Type	R	Install	Key Ack	Key MIC	Secure	Error	Request	Encryp. Key Data	R

R: Reserved

Figure 3.6. *Key information field format*

The *Key Length* field indicates the length in bytes of the PTK key: 32 bytes in the case of TKIP or 16 bytes in the case of CCMP.

The *Key Replay Counter* field is increased by one unit for each message. It enables protection against replay. The value of this field is recopied in the response to a message in which the *Key Ack* bit has been positioned at one.

The *Key Nonce* field contains the value of the random numbers assigned by the AP (ANonce) or the station (SNonce).

The *EAPOL-Key IV* field contains the initialization vector (IV) used for encrypting the data contained in the *Key Data* field.

The *Key RSC* field indicates the value of the received sequence counter (RSC) of the MAC frame. It is used after the keys have been installed.

The *Key MIC* field contains the MIC seal of the EAPOL-*Key* message.

The *Key Data Length* field indicates the size in bytes of the *Key Data* field.

The *Key Data* field contains data.

3.4.3. *Four-way handshake procedure*

The four-way handshake procedure defines four EAPOL-*Key* messages exchanged between the AP and station. This procedure enables the two end points to derive the PTK key from the PMK key and the distribution of the GTK key by the AP.

The AP sends the first message to the station if 802.1x authentication has been successful. This message contains the ANonce random number.

First EAPOL-*Key* message

Descriptor Type	254
Key Information	*Key Descriptor Version* = 1 or 2
	Key Type = 1
	Install = 0
	Key Ack = 1
	Key MIC = 0
	Secure = 0
	Error = 0
	Request = 0
	Encrypted Key Data = 0
Key Length	PTK key length
Key Replay Counter	n
Key Nonce	ANonce
EAPOL-Key IV	0
Key RSC	0
Key MIC	0
Key Data Length	22
Key Data	PMKID

Upon reception of the first message, the station verifies that the value of the *Key Replay Counter* field is greater than the value received during previous exchanges. The station generates a random number (SNonce), derives the PTK and constructs the second message containing the MIC seal calculated from the KCK key.

Second EAPOL-*Key* message

Descriptor Type	254
Key Information	*Key Descriptor Version* = 1 or 2
	Key Type = 1
	Install = 0
	Key Ack = 0
	Key MIC = 1
	Secure = 0

	Error = 0
	Request = 0
	Encrypted Key Data = 0
Key Length	0
Key Replay Counter	n
Key Nonce	SNonce
EAPOL-Key IV	0
Key RSC	0
Key MIC	MIC(KCK, EAPOL)
Key Data Length	*Key Data* field length in bytes
Key Data	RSN Information Element (IE)

Upon reception of the second message, the AP verifies that the values contained in the *Key Replay Counter* field of the first two messages are equal. It derives the PTK key, extracts the KCK key from it and checks the MIC seal contained in the *Key MIC* field.

The third message is sent by the AP to the station. It contains the GTK key encrypted with the KEK key, and an MIC seal calculated using the KCK key.

Third EAPOL-*Key* message

Descriptor Type	254
Key Information	*Key Descriptor Version* = 1 or 2
	Key Type = 1
	Install = 1
	Key Ack = 1
	Key MIC = 1
	Secure = 1
	Error = 0
	Request = 0
	Encrypted Key Data = 1
Key Length	PTK key length
Key Replay Counter	n+1
Key Nonce	ANonce
EAPOL-Key IV	(Version 2) or random value (Version 1)

Key RSC	value of sequence number to be used by the access point for MAC frames protected by the GTK key
Key MIC	MIC(KCK, EAPOL)
Key Data Length	*Key Data* field length in bytes
Key Data	RSN IE and GTK key

Upon reception of the third message, the station verifies that the value of the *Key Replay Counter* field is greater than the value received during previous exchanges. It checks that the ANonce value has not been modified and that the MIC seal value is correct.

The fourth message is sent by the station to execute the four-way handshake procedure.

Fourth EAPOL-*Key* message

Descriptor Type	254
Key Information	*Key Descriptor Version* = 1 or 2
	Key Type = 1
	Install = 0
	Key Ack = 0
	Key MIC = 1
	Secure = 1
	Error = 0
	Request = 0
	Encrypted Key Data = 0
Key Length	0
Key Replay Counter	n+1
Key Nonce	0
EAPOL-Key IV	0
Key RSC	0
Key MIC	MIC(KCK, EAPOL)
Key Data Length	0
Key Data	no data

Upon reception of the fourth message, the AP verifies that the values contained in the *Key Replay Counter* field in the third and fourth messages are equal. It checks the MIC seal contained in the *Key MIC* field.

3.4.4. *Group key handshake procedure*

The group key handshake procedure defines two EAPOL-*Key* messages exchanged between the AP and station. It takes place when the AP transmits a new GTK key to the station. The AP initializes the procedure when the station is no longer associated or authenticated. The station can start the procedure by sending an EAPOL-*Key* message with the *Request* bit positioned at one.

The first message is initialized by the AP. It sends the new GTK key encrypted with the KEK key and the MIC seal calculated using the KCK key.

First EAPOL-*Key* message

Descriptor Type	254
Key Information	*Key Descriptor Version* = 1 or 2
	Key Type = 0
	Install = 0
	Key Ack = 1
	Key MIC = 1
	Secure = 1
	Error = 0
	Request = 0
	Encrypted Key Data = 1
Key Length	0
Key Replay Counter	n+2
Key Nonce	0
EAPOL-Key IV	0 (Version 2) or random value (Version 1)
Key RSC	final sequence number in MAC frame
transmitted	
Key MIC	MIC(KCK, EAPOL)
Key Data Length	*Key Data* field length in bytes
Key Data	GTK key

Upon reception of the first message, the station verifies that the value of the *Key Replay Counter* field is greater than the value received during previous exchanges. It checks the MIC seal contained in the *Key MIC* field. It responds to the AP with the second message to satisfy the first message.

Second message EAPOL-*Key*

Descriptor Type	254
Key Information	*Key Descriptor Version* = 1 or 2
	Key Type = 0
	Install = 0
	Key Ack = 0
	Key MIC = 1
	Secure = 1
	Error = 0
	Request = 0
	Encrypted Key Data = 0
Key Length	0
Key Replay Counter	n+2
Key Nonce	0
EAPOL-Key IV	0
Key RSC	0
Key MIC	MIC(KCK, EAPOL)
Key Data Length	0
Key Data	no data

Upon reception of the second message, the AP verifies that the values contained in the *Key Replay Counter* field of the first two messages are equal. It checks the MIC seal contained in the *Key MIC* field.

3.5. WEP protocol

The WEP protocol adds 8 bytes to the MAC header (Figure 3.7):

– the WEP header is composed of IV and *KeyID* fields:

- The *IV* field, coded on 3 bytes, is the IV used to generate the pseudo-random sequence of the RC4 algorithm,

- the *KeyID* field, coded on 2 bits, enables the selection of a key from among four possible ones:

– the *Integrity Check Value (ICV)* field, coded on 4 bytes, is the result of the calculation of a cyclic redundancy check (CRC-32) applied to MAC service data unit (MSDU) data. The LLC frame constitutes the MSDU data.

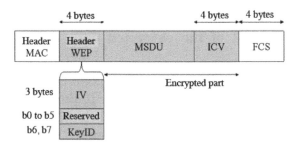

Figure 3.7. *WEP encapsulation format*

The 128-bit (or 64-bit) secret is composed of a 104-bit (or 40-bit) WEP key concatenated with the 24-bit IV. The secret determines the start sequence of the pseudo-random sequence of the RC4 algorithm (Figure 3.8).

Encryption consists of executing an exclusive OR of data including the MSDU and *ICV* fields on the one hand, and the pseudo-random sequence of the RC4 algorithm on the other hand (Figure 3.8).

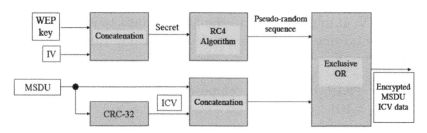

Figure 3.8. *Processing of WEP transmitted chain*

Upon reception of the MAC frame, the following operations are executed (Figure 3.9):

– the secret is reconstituted from the WEP key and the *IV* field;

– the pseudo-random sequence is initialized using the secret;

– unscrambled data (MSDU and ICV) are generated by the exclusive OR of the encrypted data and the pseudo-random sequence;

– the local calculation of the CRC-32 on the MSDU data is compared to the *ICV* field received. If the two values are equal, the MSDU data integrity check is positive. In the opposite case, the MSDU data are deleted.

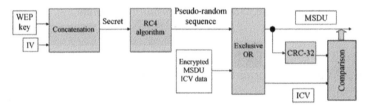

Figure 3.9. *Processing of WEP received chain*

3.6. TKIP protocol

The TKIP protocol reuses the format of the WEP protocol. It adds 4 bytes to the WEP header to introduce an extension of the IV. It adds 8 bytes to the MSDU data to join the MIC field containing the MIC seal calculated using the MICHAEL algorithm (Figure 3.10).

The TKIP header is composed of the following fields (Figure 3.10):

– the TKIP sequence counter *(TSC0)* and *TSC1* fields constitute the IV and are used during the second phase of the key mixing (key derivation) function;

– the *TSC2* to *TSC5* fields constitute an extension of the IV and are used during the first phase of the key mixing (key derivation) function;

– the *WEPseed* field is calculated from the *TSC1* field;

– the *ExtIV* field, coded on 1 byte, indicates the presence (bit set at one) of the *TSC2* to *TSC5* fields of the IV extension;

– as for the WEP protocol, the *KeyID* field, coded on 2 bits, enables the selection of one key from among four possible ones.

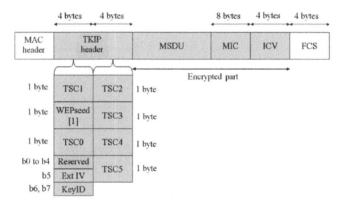

Figure 3.10. *TKIP encapsulation format*

The processing of the transmitted chain is shown in Figure 3.11.

The MIC seal is calculated using the MAC addresses of the source and the destination; the priority byte, and the MSDU data. The priority byte contains the priority level of the frame.

The set composed of MSDU and MIC data can be fragmented. In this case, the IV is increased by one unit for each fragment. Conversely, the IV extension keeps the same value for all fragments of a single MSDU.

For each MSDU, two key mixing (key derivation) phases are used to calculate the secret used for WEP processing:

– the first phase operates using the Transmit Address (TA), the TK key and the TSC2 to TSC5 vectors;

– the second phase operates using the TKIP-mixed Transmit Address And Key (TTAK) key, the TK key and the TSC0 and TSC1 vectors. The TTAK key constitutes an intermediary key produced during phase one.

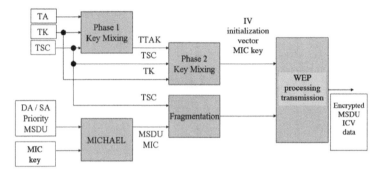

Figure 3.11. *Processing of the TKIP transmitted chain*

The processing of the received chain is shown in Figure 3.12.

The receiver extracts the *TSC* fields from the TKIP header and verifies the sequencing in order to protect itself from replays.

The combination of the *TSC* fields with the TK and the MAC address of the TA generator enables the IV and the RC4 key to be reconstituted for the WEP decryption.

If WEP processing indicates a positive check from the *ICV* field, the fragments are reassembled.

The result of the MIC seal calculation using the MIC key and unscrambled MSDU data is compared to the value of the

MIC field received. If the two values match, the integrity check is positive and the MSDU data are accepted.

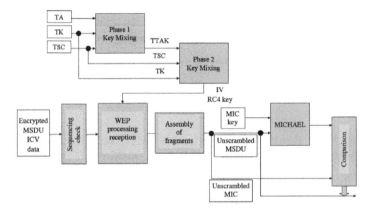

Figure 3.12. *Processing of the TKIP received chain*

3.7. CCMP protocol

The CCMP protocol adds 16 bytes to the MAC frame (Figure 3.13):

– 8 bytes for the CCMP header;

– 8 bytes for the MIC seal.

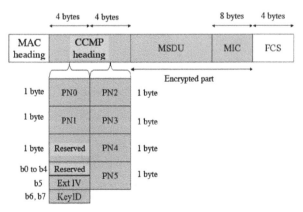

Figure 3.13. *CCMP encapsulation format*

The CCMP header resembles the TKIP protocol header. It is constructed from the number of the Packet Number (PN) frame; the *ExtIV* field, coded on 1 bit and indicating the presence (bit set at one) of the *PN2* to *PN5* fields; and the *KeyID* field, coded on 2 bits, used to select a key from among four possible keys.

The processing of the transmitted chain is shown in Figure 3.14.

The AES algorithm provides the MIC seal and encryption of the MSDU and MIC data. It is supplied by the following values:

– the Additional Authentication Data (AAD) parameter, built from the MAC header, with the exception of fields that can be modified during a retransmission (for example, the *Duration* field);

– the Nonce parameter, built from the priority byte; the second address contained in the MAC header (A2); and the PN frame number. The value of the *PN* field is increased by one unit for each frame generated;

– the TK key.

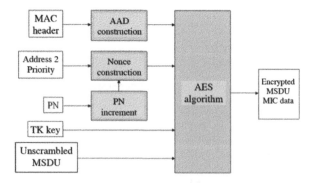

Figure 3.14. *Processing of CCMP transmitted chain*

The processing of the CCMP received chain is shown in Figure 3.15.

The AES algorithm is used to reproduce unscrambled MSDU data. It is supplied by the following values:

– the AAD parameter;

– the Nonce parameter;

– the MIC field to execute an integrity check;

– the TK key.

A check on the *PN* field enables protection from replays.

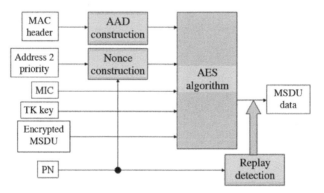

Figure 3.15. *Processing of the CCMP received chain*

4

IPSec Mechanism

4.1. Review of IP protocols

The Internet Protocol (IP) is used by routers to transfer packets between the two hosts, the source and destination. To do this, the routers use the destination address of the IP header and consult a routing table, which provides an output interface depending on the destination IP address.

The IP packet is generated by the source. Its size is defined by the level 2 protocol used by the network to which the host is connected. For example, Ethernet limits the size of the packet or Maximum Transmission Unit (MTU) to 1,500 bytes.

The IP protocol provides an unreliable packet transfer service. The router neither acknowledges data received nor checks flows or errors in the IP packet.

4.1.1. *IPv4 protocol*

The header of the IPv4 protocol contains the following fields (Figure 4.1):

Version: this field, coded on 4 bits, includes the version number of the IP protocol. It has a value of 4.

Internet Header Length (*IHL*): this field, coded on 4 bits, includes the size of the IP header in multiples of 4 bytes. For an optionless header, this field has a value of 5.

Type of Service (*ToS*): this field, coded on 1 byte, was defined in the original version of the standard to implement priority management and give information on the level required for delay, output, reliability and cost.

The *ToS* field is replaced by the DiffServ Code Point (*DSCP*) field coded on 6 bits and the Explicit Congestion Notification (*ECN*) field coded on 2 bits.

DSCP marking is used to build service quality in IP networks. Routers use marking to select Per-Hop Behavior (PHB) corresponding to a specific process for data transfer.

The ECN mechanism relies on the use of fields in the IP and Transmission Control Protocol (TCP) protocol headers to alert first the destination station, and then the source station if congestion develops, causing reduced rate from the source.

Total Length: this field, coded on 2 bytes, contains the size of the packet, including the IPv4 header and encapsulated data. It authorizes a packet length of 65,535 bytes.

Such a length is impractical; however, it is generally limited by the level 2 protocol.

The router ensures the interconnection of networks, the connectivity function of which is ensured by technology different to IP, such as Ethernet for example. Therefore, it enables a change of networks; that is, a modification of the physical layer (level 1 layer) and the data link layer (level 2 layer).

The transfer executed by the router can send the packet onto a new network, the MTU value of which is less than the original one. In this case, the router executes a fragmentation of the IPv4 packet in order to adapt itself to the new MTU value. The reassembly of the fragments to recreate the original packet is done by the destination. The MTU has a minimum value of 576 bytes.

Identification: this field, coded on 2 bytes, contains a value enabling the reassembly on reception of the IP packet fragments.

Flags: this field contains three bits used in the following way:

– the first bit is always positioned at zero;

– the second Don't Fragment (DF) bit enables the authorization (bit at zero) or forbidding (bit atone) of fragmentation;

– the Third More Fragment (MF) bit identifies the last fragment (bit at zero) or intermediary fragments (bit at one).

Fragment Offset: this field, coded on 12 bits, indicates the positioning of the fragment in the initial packet. The measurement is done in multiples of 8 bytes. The value of the first fragment field is 0.

Time to Live (*TTL*): this field, coded on 1 byte, contains the maximum number of routers crossed by the packet. Each router crossed decreases the value of this field by one unit. When the value reaches 0, the packet is deleted, and an Internet Control Message Protocol (ICMPv4) error message is sent to the source.

Protocol: this field, coded on 1 byte, contains a value used to identify the type of data encapsulated by the IPv4 header (Table 4.1).

Protocol field	Data encapsulated
1	ICMPv4 (Internet Control Message Protocol) message
2	IGMP (Internet Group Management Protocol) message
4	IPv4 packet
6	TCP (transmission control protocol) segment
17	UDP (user datagram protocol) segment
41	IPv6 packet
50	ESP (encapsulating security payload) extension
51	AH (authentication header) extension
89	OSPFv2 (open shortest path first) message

Table 4.1. *Protocol field values*

Header Checksum: this field, coded on 2 bytes, contains a checksum calculated only on the header.

Source Address: this field, coded on 4 bytes, contains the IPv4 address of the packet source.

Destination Address: this field, coded on 4 bytes, contains the IPv4 address of the packet destination.

Theoretically, the IPv4 header can contain zero, one or several options. The options initially defined are no longer used. Conversely, a new option, *Router Alert*, was subsequently introduced to force the router to analyze encapsulated data, even though it is not the destination. This option is used in association with certain ReSerVation Protocol (RSVP) messages.

4.1.2. *IPv6 protocol*

The IP version 6 (IPv6) protocol is the new protocol intended to succeed the IPv4 protocol. The principal modification contributed by the IPv6 protocol concerns the size (16 bytes) allocated to the source and destination addresses, which results in a larger header (40 bytes).

```
0                   1                   2                   3
0 1 2 3 4 5 6 7 8 9 0 1 2 3 4 5 6 7 8 9 0 1 2 3 4 5 6 7 8 9 0 1
+-+-+-+-+-+-+-+-+-+-+-+-+-+-+-+-+-+-+-+-+-+-+-+-+-+-+-+-+-+-+-+-+
|Version|  IHL  |Type of Service|          Total Length         |
+-+-+-+-+-+-+-+-+-+-+-+-+-+-+-+-+-+-+-+-+-+-+-+-+-+-+-+-+-+-+-+-+
|         Identification        |Flags|      Fragment Offset    |
+-+-+-+-+-+-+-+-+-+-+-+-+-+-+-+-+-+-+-+-+-+-+-+-+-+-+-+-+-+-+-+-+
|  Time To Live |    Protocol   |        Header Checksum         |
+-+-+-+-+-+-+-+-+-+-+-+-+-+-+-+-+-+-+-+-+-+-+-+-+-+-+-+-+-+-+-+-+
|                        Source Address                         |
+-+-+-+-+-+-+-+-+-+-+-+-+-+-+-+-+-+-+-+-+-+-+-+-+-+-+-+-+-+-+-+-+
|                      Destination Address                      |
+-+-+-+-+-+-+-+-+-+-+-+-+-+-+-+-+-+-+-+-+-+-+-+-+-+-+-+-+-+-+-+-+
```

Figure 4.1. *IPv4 header format*

The IPv6 protocol has introduced simplifications compared to the IPv4 protocol, enabling better performance with regard to packet processing by routers. The following fields have been removed from the IPv6 header:

– *IHL*, since the IPv6 header has a fixed length of 40 bytes;

– *Identification*, *Flags*, *Fragment Offset*, since fragmentation is the subject of an extension;

– *Header Checksum*, since transmission is considered to be of good quality, and binary errors are rare.

Options have been removed from the basic header and replaced by new headers called extensions. Other than the *Hop-by-Hop* option, which is processed by all intermediary routers, the other options are only taken into account by the destinations.

Fragmentation of the IPv4 packet is executed by the router, which is not the case in IPv6. When an IPv6 packet is sent, the source must discover the minimum MTU value. For applications that cannot segment the message, fragmentation of the IPv6 packet is executed by the source. This provision improves router performance.

The header of the IPv6 protocol contains the following fields (Figure 4.2):

Version: this field, coded on 4 bits, displays the version number of the IP protocol. It has a value of 6;

Traffic Class: this field, coded on 1 byte, is equivalent to the *DSCP* and *ECN* fields in the IPv4 header;

Flow Label: this field, coded on 20 bits, displays a unique number chosen by the source. Its purpose is to facilitate the work of routers to implement classification functions for a flow;

Payload Length: this field, coded on 2 bytes, displays the size of the data encapsulated by the IP header. It is different from the *Total Length* field in the IPv4 header, which indicates the packet size;

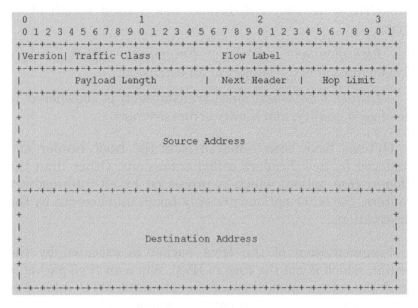

Figure 4.2. *IPv6 header format*

Next Header: this field, coded on 1 byte, displays the identifier of the next header of data encapsulated by the IP header. This field plays the same role as the *Protocol* field in the IPv4 header. New values for this field have been defined to take extensions into account (Table 4.2);

Next header field	Data encapsulated
0	*Hop-by-Hop* extension
4	IPv4 packet
6	TCP (transmission control protocol) segment
17	UDP (user datagram protocol) segment
41	IPv6 packet
43	*Routing* extension
44	*Fragment* extension
50	ESP (encapsulating security payload) extension
51	AH (authentication header) extension
58	ICMPv6 (Internet Control Message Protocol) message
59	End of extensions
60	*Destination* extension
89	OSPFv3 (open shortest path first) message

Table 4.2. *Next Header field values*

Hop Limit: this field, coded on 1 byte, is equivalent to the *TTL* field in the IPv4 header;

Source Address: this field, coded on 16 bytes, shows the IP address of the packet source;

Destination Address: this field, coded on 16 bytes, shows the IP address of the packet destination.

4.2. IPSec architecture

Security services (authentication, integrity and confidentiality) are offered in an identical way in IPv4 and IPv6. Their implementation is optional in IPv4 and mandatory in IPv6. Their use is optional.

The Internet Protocol Security (IPSec) mechanism can be used with LAN hosts or security gateways located at the interface between the LAN and WAN networks.

The first application consists of linking remote LAN networks via an untrusted WAN network. The IPSec mechanism is deployed in the security gateways of each LAN network. In this case, the two LAN networks are considered protected networks (Figure 4.3).

The second application consists of linking a host to a remote LAN network via a LAN network (the host's) and an untrusted WAN network. The IPSec mechanism is deployed in the host and a security gateway located in the remote LAN network, considered a protected network (Figure 4.3).

The third application consists of linking two hosts via two LAN networks and an untrusted WAN network. The IPSec mechanism is deployed from end-to-end, in each host (Figure 4.3).

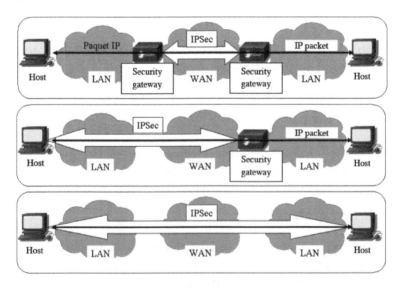

Figure 4.3. *IPSec architecture*

4.2.1. *Security headers*

The IPSec mechanism introduces two IPv4 or IPv6 header extensions:

– The Authentication Header (AH) is designed to ensure the integrity and authentication of IP packets without data encryption (no confidentiality).

– The Encapsulating Security Payload (ESP) ensures the integrity, authentication and confidentiality of IP packets.

4.2.1.1. *AH extension*

The AH extension offers authentication and data integrity services and enables protection against IP packet replay. The same AH extension is used in association with an IPv4 or IPv6 header. The presence of the extension is indicated by the *Next Header* (in IPv6) or *Protocol* (in IPv4) field of the previous header, with a value of 51.

In addition to the *Next Header* field, the AH extension contains the following fields (Figure 4.4):

Payload Length: this field, coded on 1 byte, provides the extension size in multiples of 4 bytes, not including the first 8 bytes. The size of the extension in IPv6 must remain a multiple of 8 bytes.

Security Parameters Index (*SPI*): this field, coded on 4 bytes, contains a value pertaining to the previously negotiated Security Association (SA).

Sequence Number: this field, coded on 4 bytes, contains a value increased by one unit for each IPv4 or IPv6 packet transmitted. This field enables protection against replay. This field has a value of 1 for the first packet transmitted. When the counter reaches the maximum value, a new SA must be negotiated in order to avoid the start of a new cycle.

An Extended Sequence Number (ESN) coded on 8 bytes constitutes an option making it possible for the lifetime of the SA to be prolonged. In order to preserve the structure of the extension, the 32 least significant bits are transmitted in the *Sequence Number* field. However, the data digest is calculated on all 64 bits.

Integrity Check Value (*ICV*): this field is coded on a multiple of 4 bytes and contains the data seal ensuring authentication and integrity checking.

```
0                   1                   2                   3
0 1 2 3 4 5 6 7 8 9 0 1 2 3 4 5 6 7 8 9 0 1 2 3 4 5 6 7 8 9 0 1
+-+-+-+-+-+-+-+-+-+-+-+-+-+-+-+-+-+-+-+-+-+-+-+-+-+-+-+-+-+-+-+-+
| Next Header  |  Payload Len  |            RESERVED           |
+-+-+-+-+-+-+-+-+-+-+-+-+-+-+-+-+-+-+-+-+-+-+-+-+-+-+-+-+-+-+-+-+
|                 Security Parameters Index (SPI)              |
+-+-+-+-+-+-+-+-+-+-+-+-+-+-+-+-+-+-+-+-+-+-+-+-+-+-+-+-+-+-+-+-+
|                    Sequence Number Field                     |
+-+-+-+-+-+-+-+-+-+-+-+-+-+-+-+-+-+-+-+-+-+-+-+-+-+-+-+-+-+-+-+-+
|                                                              |
+              Integrity Check Value-ICV (variable)           |
|                                                              |
+-+-+-+-+-+-+-+-+-+-+-+-+-+-+-+-+-+-+-+-+-+-+-+-+-+-+-+-+-+-+-+-+
```

Figure 4.4. *AH extension format*

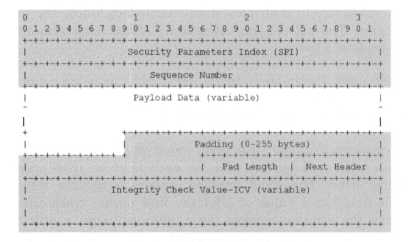

Figure 4.5. *ESP extension format*

4.2.1.2. *ESP extension*

The ESP extension provides a confidentiality service in addition to those offered by the AH extension. The same ESP extension is used in association with an IPv4 or IPv6 header. The presence of the extension is indicated by the *Next Header* (in IPv6) or *Protocol* (in IPv4) field of the previous header, with a value of 50.

The ESP extension contains the same fields as the AH extension. It starts with the *SPI* and *Sequence Number* fields (Figure 4.5). After these fields comes the encapsulated data, which may contain synchronization data of the initialization vector (IV) encryptor. Following the encapsulated data, the extension ends with the following fields: *Padding*, *Pad Length*, *Next Header* and optionally *ICV* (authentication data).

The *Padding* field is necessary when block encryption is used, and the block must be of a certain size, and to align the packet size with a multiple of 4 bytes.

4.2.1.3. *Modes*

For transport mode, the AH or ESP header is inserted between the IP header and the source IP packet payload. In the IPv6 environment, the AH or ESP header appears after the *Hop-by-Hop*, *Destination*, *Routing*, and *Fragment* extensions.

For tunnel mode, the AH or ESP header encapsulates the source IP packet, and the whole is encapsulated in its turn by a new IP header. The tunnel corresponds to a data structure in which an IP packet contains another IP packet.

When the AH header is used, authentication is applied to the whole packet except the variable fields of the IP header (Figure 4.6).

The variable fields of the IPv4 header are set at zero to calculate the authentication digest:

– *DSCP*: the value of this field can be modified by an intermediary router when it checks traffic characteristics;

– *ECN*: the value of this field can be modified by an intermediary router to alert the destination that congestion is developing;

– *DF*: this bit can be set at one by an intermediary router;

– *Fragment Offset*: insertion of the AH header occurs on non-fragmented IP packets, and therefore this field has a value of zero;

– *TTL*: the value of this field is decreased by one unit per each router crossed;

– *Checksum*: the value of this field is recalculated as soon as a field in the IP header changes value.

The group of IPv4 header options is considered as a single entity. Some options can be modified by an intermediary router. If a single modifiable option appears, the group of options is set at zero for the authentication digest calculation.

The variable fields of the IPv6 header are set at zero for the authentication digest calculation. These are identical fields to the ones in the IPv4 header (*DSCP*, *ECN* and *Hop Limit*), as well as the *Flow Label* field.

The *Hop-by-Hop* and *Destination* extensions of the IPv6 header have a bit that indicates whether the option can be modified by an intermediary router or not. If this bit is set at one, the extension is set at zero for the authentication digest calculation.

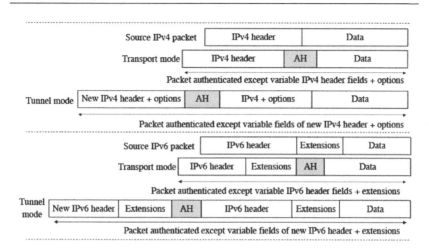

Figure 4.6. *Position of AH header*

When the ESP header is used in transport mode, the confidentiality service is applied to the encapsulated data and ESP tail. Authentication and integrity services cover the ESP header, encapsulated data and ESP tail (Figure 4.7).

When the ESP header is used in tunnel mode, the confidentiality service is applied to the source IP packet and ESP tail. Authentication and integrity services cover the ESP header, source IP packet, and ESP tail (Figure 4.7).

Note that in transport mode, the *Destination* extension can appear before, after or simultaneously before and after the AH or ESP extension.

4.2.2. *Security association*

A SA is a simple connection between two end points offering security services (confidentiality, integrity and authentication) to traffic. Security services are provided by the use of AH or ESP extensions. To secure a two-directional communication between two end points, an SA pair is required. The Internet Key Exchange (IKE) protocol dynamically ensures the creation of the SA.

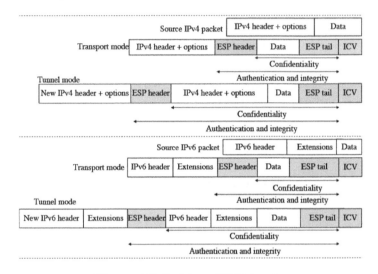

Figure 4.7. *Position of ESP header*

An SA contains the following parameters:

– the authentication algorithm and the key in order to generate the AH extension;

– the encryption algorithm and the key in order to generate the ESP extension;

– the authentication algorithm and the key in order to generate the ESP extension, if this service is used;

– the lifetime of the SA;

– the encapsulation mode (tunnel or transport).

The IPSec mechanism defines three databases:

– Security Policy Database (SPD). This defines the security policy to be applied to input and output traffic for a host or a security gateway;

– Security Association Database (SAD). This contains the parameters applied to an SA;

– Peer Authorization Database (PAD). This provides a link between the IKEv2 protocol and SPD database.

The selector is the mechanism enabling identification at the source of the SA to be applied to traffic. The selector uses fields (*Protocol* or *Next Header* and source or destination IP address) of the IP headers and fields (source or destination port) of the TCP or User Datagram Protocol (UDP) headers.

When a packet originates from the interface of the protected network (outgoing packet), the selector makes it possible to obtain entry to the SPD database that defines the process to be applied to the packet:

– BYPASS: the packet is transmitted without a security service;

– DISCARD: the packet is deleted;

– PROTECT: the security service is applied to the packet. If the SA is not established, the IKE protocol is invoked. If the SA exists, the database returns a pointer to the SAD database.

The deletion of a packet causes the generation from the security gateway toward the source of an ICMP message with the following characteristics:

– in the IPv4 environment, Type = 3 (destination unreachable) and Code = 13 (Communication Administratively Prohibited);

– in the IPv6 environment, Type = 1 (destination unreachable) and Code = 1 (Communication with Destination Administratively Prohibited).

The Security Parameter Index (SPI) is a field in the AH or ESP header used by the destination to identify the SA in a unique way. The destination uses this index to extract the SA from the SAD database.

When the packet originates from the interface of the non-protected network (incoming packet), the SPD database is consulted if the packet is not protected, and the instruction (BYPASS or DISCARD) is applied. If the IP packet is protected, the SPI field is used to recover the SA.

4.2.3. *PMTU processing*

The introduction of ESP or AH extensions has an impact on the packet size, which can exceed the MTU value. The Path MTU (PMTU) mechanism used to find the MTU value on the route makes it possible to avoid packet fragmentation by intermediary routers.

In the IPv4 environment, the PMTU mechanism uses the DF bit in the IP header positioned at one. In the tunnel mode used between two security gateways, it is, therefore, necessary for the security gateway to copy the value of this bit in the new header.

The PMTU mechanism uses an ICMP message indicating the MTU value. This message is sent back by an intermediary router, with the following characteristics:

– in the IPv4 environment, Type = 3 (Destination Unreachable) and Code = 4 (Fragmentation needed and DF set);

– in the IPv6 environment, *Type = 2 (Packet Too Big)* and *Code = 0 (Fragmentation needed)*.

Thus, the security gateway can receive this message from an unauthenticated source (an intermediary router) in the SA. It must then analyze the content of the ICMP message to recover the information that will grant it access to the SAD database (IP addresses and SPI field). It must resend an ICMP message to the host with a new PMTU value taking the size of the AH or ESP extensions into account.

4.3. IKEv2 protocol

The IKEv2 protocol is more simplified than the previous version. It combines the functionalities defined in IKEv1 and Internet Security Association And Key Management Protocol (ISAKMP) while removing unnecessary processes. It eliminates the generic character of the previous version, integrating the domain of interpretation DOI function, which defines the parameters specific to the ESP/AH SA.

Each IKEv2 message is composed of an HDR header and a sequence of blocks. The IKEv2 message is encapsulated by a UDP header with source and destination port values of 500 or 4500. When the 4500 port is used, the IKEv2 message is preceded by 4 bytes at zero.

4.3.1. *Message header*

The header of the IKE message contains the following fields (Figure 4.8):

Initiator's SPI: this field, coded on 8 bytes, incorporates a value chosen by the initiator. It initializes identification of the IKE SA.

Responder's SPI: this field, coded on 8 bytes, incorporates a value chosen by the responder. It completes the identification of the IKE SA.

Next Payload: this field, coded on 1 byte, incorporates the indication of the type of block following the header (Table 4.3).

Notation	Designation
SA	*Security Association*
KE	*Key Exchange*
IDi	*Identification – initiator*
IDr	*Identification – responder*
CERT	*Certificate*
CERTREQ	*Certificate Request*
AUTH	*Authentication*
Ni	*Nonce – initiator*
Nr	*Nonce – responder*
N	*Notification*
D	*Delete*
V	*Vendor ID*
TSi	*Traffic Selector – initiator*
TSr	*Traffic Selector – responder*
SK	*Encrypted and Authenticated*
CP	*Configuration*
EAP	*Extensible Authentication*

Table 4.3. *Block types*

Major Version: this field, coded on 4 bits, indicates the maximum value of the IKE protocol version that can be used. This value is equal to 2 for the implementation of the IKEv2 protocol.

Minor Version: this field, coded on 4 bits, indicates the minimum value of the IKE protocol version. This value is equal to 0 for the implementation of the IKEv2 protocol.

Exchange Type: this field, coded on 1 byte, indicates the type of exchange to which the message belongs:

– IKE_SA_INIT: this exchange concerns the first phase of the establishment of the IKE SA.

– IKE_AUTH: this exchange concerns the second phase of the establishment of the IKE SA.

– CREATE_CHILD_SA: this exchange concerns the establishment of the ESP/AH SA.

– INFORMATIONAL: this exchange concerns event notification.

Each type of exchange imposes a certain number of required blocks composing the message and defines optional blocks.

Flags: this field includes the following three flags:

– R (response): this flag, positioned at one, indicates that this message is a response. An IKE termination must not respond to a response except when authentication has failed.

– V (version): this flag, positioned at one, indicates that the IKE termination is able to process a version higher than the one shown in the *Major Version* field.

– I (initiator): this flag, positioned at one, indicates that the message is generated by the initiator of the IKE SA.

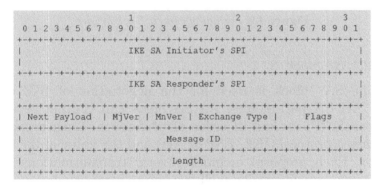

Figure 4.8. *IKE message header format*

Message ID: this field, coded on 4 bytes, is an identifier used to control the retransmission of lost messages and to correlate the request and response. It also protects against replay attacks.

Length: this field, coded on 4 bytes, includes the IKE message size.

4.3.2. *Blocks*

Each block starts with a generic header containing the *Next Payload* field, the C (critical) bit and the *Payload Length* field (Figure 4.9). The *Next Payload* field indicates the type of block that comes next, thus enabling chaining. The C bit determines the process to be executed when the receiver does not recognize the block:

– if the C bit is positioned at one, the receiver rejects the message;

– if the C bit is positioned at zero, the receiver ignores the block and processes the next block.

```
 0                   1                   2                   3
 0 1 2 3 4 5 6 7 8 9 0 1 2 3 4 5 6 7 8 9 0 1 2 3 4 5 6 7 8 9 0 1
+-+-+-+-+-+-+-+-+-+-+-+-+-+-+-+-+-+-+-+-+-+-+-+-+-+-+-+-+-+-+-+-+
| Next Payload |C|  Reserved   |         Payload Length         |
+-+-+-+-+-+-+-+-+-+-+-+-+-+-+-+-+-+-+-+-+-+-+-+-+-+-+-+-+-+-+-+-+
```

Figure 4.9. *Format of generic block header*

4.3.2.1. *SA block*

The SA block is used for the negotiation of IKE and ESP/AH SA parameters. An SA block can contain several proposals (P) ranked in order of preference. Each proposal defines a protocol (IKE, ESP or AH) and the value of the SPI index. Each proposal includes several transformations (T), and each transformation includes one or more attributes (A).

The transformation T involves the following operations:

– the encryption algorithm ENCR. This operation is used for negotiation pertaining to the IKE and ESP protocols.

ENCR_DES_IV64
ENCR_DES 2
ENCR_3DES
ENCR_RC5
ENCR_IDEA

ENCR_CAST
ENCR_BLOWFISH
ENCR_3IDEA
ENCR_DES_IV32
ENCR_NULL
ENCR_AES_CBC
ENCR_AES_CTR

– the pseudo-random function PRF. This operation is used for negotiation pertaining to the IKE protocol.

PRF_HMAC_MD5
PRF_HMAC_SHA1
PRF_HMAC_TIGER

– the integrity algorithm INTEG. This operation is used for negotiation pertaining to IKE, AH and (optionally) ESP protocols.

NONE
AUTH_HMAC_MD5_96
AUTH_HMAC_SHA1_96
AUTH_DES_MAC
AUTH_KPDK_MD5
AUTH_AES_XCBC_96

– the Diffie–Hellman (D-H) group. This operation is used for negotiation pertaining to IKE and (optionally) AH and ESP protocols.

NONE
768-bit MODP
1024-bit MODP
1536-bit MODP
2048-bit MODP
3072-bit MODP
4096-bit MODP
6144-bit MODP
8192-bit MODP

– the ESN. This operation is used for negotiation pertaining to the AH and ESP protocols.

No ESNs
ESNs

The attribute A specifies the length of the encryption algorithm key defined in the ENCR transformation. The other transformations, PRF, INTEG, D-H and ESN have no attributes.

4.3.2.2. *KE block*

The Key Exchange (KE) block contains the public Diffie–Hellman value, enabling each end point (initiator and responder) to construct a shared secret. The block also mentions the D-H group defined in the SA block.

4.3.2.3. *IDi and IDr blocks*

Identification initiator (IDi) and Identification responder (IDr) blocks contain an identification of the initiator of the IKE message and the responder. This identification is based on an IPv4 or IPv6 address, a name, a messaging address or a group of bytes.

ID_IPV4_ADDR
ID_FQDN
ID_RFC822_ADDR
ID_IPV6_ADDR
ID_DER_ASN1_DN
ID_DER_ASN1_GN
ID_KEY_ID

4.3.2.4. *CERT block*

The certificate (CERT) block provides a means of transporting a certificate or information pertaining to authentication.

PKCS #7 wrapped X.509 certificate
PGP certificate
DNS signed key

X.509 certificate – signature

Kerberos token

Certificate revocation list

Authority revocation list

SPKI certificate

X.509 certificate – attribute

Raw RSA key

Hash and URL of X.509 certificate

Hash and URL of X.509 bundle

4.3.2.5. *CERTREQ block*

The certificate request (CERTREQ) block is a request pertaining to a certificate. It is used in the response of the IKE_INIT_SA exchange response or in the IKE_AUTH exchange request. It also indicates the certification authority for the required certificate.

4.3.2.6. *The AUTH block*

The authentication (AUTH) block contains the authentication digest of the message. The block also specifies the method used.

RSA digital signature

Shared key message integrity code

DSS digital signature

4.3.2.7. *Ni and Nr blocks*

Nonce initiator (Ni) and nonce responder (Nr) blocks contain a random number generated by the initiator and responder. These numbers are used in the creation of derived keys.

4.3.2.8. *N block*

The notification (N) blocks contain error messages indicating the reason why the SA cannot be established.

UNSUPPORTED_CRITICAL_PAYLOAD
INVALID_IKE_SPI
INVALID_MAJOR_VERSION
INVALID_SYNTAX
INVALID_MESSAGE_ID
INVALID_SPI
NO_PROPOSAL_CHOSEN
INVALID_KE_PAYLOAD
AUTHENTICATION_FAILED
SINGLE_PAIR_REQUIRED
NO_ADDITIONAL_SAS
INTERNAL_ADDRESS_FAILURE
FAILED_CP_REQUIRED
TS_UNACCEPTABLE
INVALID_SELECTORS
TEMPORARY_FAILURE
CHILD_SA_NOT_FOUND

The N block also contains status messages that an SA management process wishes to communicate to a remote process.

INITIAL_CONTACT
SET_WINDOW_SIZE
ADDITIONAL_TS_POSSIBLE
IPCOMP_SUPPORTED
NAT_DETECTION_SOURCE_IP
NAT_DETECTION_DESTINATION_IP
COOKIE
USE_TRANSPORT_MODE
HTTP_CERT_LOOKUP_SUPPORTED
REKEY_SA
ESP_TFC_PADDING_NOT_SUPPORTED
NON_FIRST_FRAGMENTS_ALSO

4.3.2.9. *D block*

The Delete (D) block includes the SPI index of the SA that the message source wishes to delete. For an AH or ESP protocol, it is possible to specify several index values. It is also possible to string several D blocks together in a single IKEv2 message.

4.3.2.10. *V block*

The Vendor ID (V) block announces that the message source is capable of accepting private extensions of the IKEv2 protocol. These extensions can involve the introduction of new blocks, new types of exchange or new notification information.

4.3.2.11. *TS block*

The Traffic Selector (TS) block identifies the flows for which the ESP/AH SA is implemented. Flow determination is based on the following information:

– the type of data encapsulated by the IP header, stated in the *Protocol* fields of the IPv4 header or the *Next Header* field of the IPv6 header;

– the range of source and destination IP addresses;

– the range of source and destination port numbers if the IP header encapsulates UDP or TCP segments;

– the ICMP message type and code.

4.3.2.12. *SK block*

The SK (encrypted and authenticated) block is always located at the end of the IKEv2 message. The encryption and integrity algorithms of the IKEv2 algorithms are negotiated during the implementation of the IKEv2 SA.

4.3.2.13. *CP block*

The Configuration (CP) block is used to exchange configuration information between the two end points. In the case of an ESP/AH SA between a host and a security gateway, the host can request information concerning a host in the protected network.

INTERNAL_IP4_ADDRESS
INTERNAL_IP4_NETMASK
INTERNAL_IP4_DNS
INTERNAL_IP4_NBNS
INTERNAL_IP4_DHCP
APPLICATION_VERSION
INTERNAL_IP6_ADDRESS
INTERNAL_IP6_DNS
INTERNAL_IP6_DHCP
INTERNAL_IP4_SUBNET
SUPPORTED_ATTRIBUTES
INTERNAL_IP6_SUBNET

4.3.2.14. *EAP block*

The Extensible Authentication Protocol (EAP) block enables authentication of the IKE SA by the EAP protocol.

4.3.3. Procedure

4.3.3.1. *IKE_SA_INIT exchange*

The first exchange, IKE_SA_INIT, negotiates cryptographic algorithms and random numbers and executes a Diffie–Hellman exchange in order to create an IKE SA.

The initiator generates an IKE message containing the HDR header and the SAi1, KEi and Ni blocks. The HDR header contains the initiator's SPI index, version numbers and flags. The SAi1 block contains the cryptographic algorithms proposed by the initiator for the IKE SA. The KEi block includes the Diffie–Hellman group and public value. The Ni block displays the initiator's random number (Figure 4.10).

The responder chooses a cryptograph series from the initiator's proposals and includes it in the SAr1 block. It completes the exchange of Diffie–Hellman keys with the KEr block. It sends its random number in the Nr block (Figure 4.10). It can possibly communicate a list of certificate authorities in the CERTREQ block.

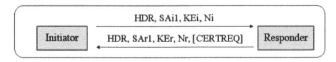

Figure 4.10. *IKE_SA_INIT exchange*

At this stage of the negotiation, each end point can generate the SKEYSEED key. The keys used for encryption and integrity of IKE messages are produced by the SKEYSEED key and are known as SK_e (encryption) and SK_a (integrity). Message protection involves only the blocks; the header is not included.

The two different directions of traffic use different keys. The keys used to protect messages from the initiator are SK_ai and SK_ei. The keys used to protect messages in the other direction are SK_ar and SK_er.

Other keys are also derived from the SKEYSEED key. The SK_d key is used to derive the keys used in the ESP/AH SA. The SK_p key is used to calculate the AUTH block digest.

The SKEYSEED key is calculated using the Diffie–Hellman secret (D-H key) and the random numbers Ni and Nr:

SKEYSEED = PRF (D-H key, Ni | Nr)

The keys SK_d, SK_ai, SK_ar, SK_ei, SK_er, SK_pi and SK_pr are generated as follows:

SK_d = PRF (SKEYSEED, Ni | Nr | SPIi | SPIr | 0x01)
SK_ai = PRF (SKEYSEED, SK_d | Ni | Nr | SPIi | SPIr | 0x02)
SK_ar = PRF (SKEYSEED, SK_ai | Ni | Nr | SPIi | SPIr | 0x03)
SK_ei = PRF (SKEYSEED, SK_ar | Ni | Nr | SPIi | SPIr | 0x04)
SK_er = PRF (SKEYSEED, SK_ei | Ni | Nr | SPIi | SPIr | 0x05)
SK_pi = PRF (SKEYSEED, SK_er | Ni | Nr | SPIi | SPIr | 0x06)
SK_pr = PRF (SKEYSEED, SK_pi | Ni | Nr | SPIi | SPIr | 0x07)

4.3.3.2. *IKE_AUTH exchange*

The second exchange, IKE_AUTH, is used to authenticate previous IKE messages and communicate identities and possibly exchange certificates, as well as to establish the first AH/ESP SA. These messages are completely encrypted and protected by the keys established during the IKE_SA_INIT exchange.

The initiator indicates its identity with the IDi block. It authenticates its identity and protects the integrity of the first message in the IKE_SA_INIT exchange using the AUTH block. It can possibly send its certificate in the CERT block and the certification authority's identity in the useful CERTREQ payload load (Figure 4.11).

The optional IDr block enables the initiator to specify which of the responder's identities it wishes to communicate with (Figure 4.11).

The initiator starts the AH/ESP SA negotiation with the SAi2 block. The TSi block specifies the characteristics of the packets transferred by the initiator. The TSr block specifies the address for packets transferred to the responder (Figure 4.11).

The notation SK{ ... } indicates that the blocks are completely encrypted and protected (Figure 4.11).

The responder communicates its identity in the IDr block. It may send a certificate. It authenticates its identity and protects the integrity of the second message of the IKE_SA_INIT exchange. It completes the negotiation of the ESP/AH SA with the SAr2 block (Figure 4.11).

Figure 4.11. *IKE_AUTH exchange*

4.3.3.3. *CREATE_CHILD_SA exchange*

The CREATE_CHILD_SA exchange is used to create the ESP/AH SA and to renew IKE and ESP/AH SA keys.

For the exchange involving the creation of the ESP/AH SA, the initiator finalizes the SA in the SA block and the traffic selectors proposed for the SA in the TSi and TSr blocks. It transmits a random number in the Ni block and optionally a Diffie–Hellman value in the KEi block (Figure 4.12).

The responder confirms the SA offer in the SA block. It transmits a Diffie–Hellman value in the KEr block, if the KEi block has been included in the request (Figure 4.12).

Figure 4.12. *CREATE_CHILD_SA exchange creation of ESP/AH SA*

The KEYMAT key, used for the ESP/AH SA, is derived from the SK_d key and the random numbers Ni and Nr. If the exchange contains Diffie–Hellman values, the secret D-H key obtained also participates in the creation of the KEYMAT key.

$$KEYMAT = PRF (SK\_d, Ni \mid Nr)$$

$$KEYMAT = PRF (SK\_d, D\text{-}H \ key \mid Ni \mid Nr)$$

To renew the IKE SA key, the initiator sends the SA in the SA block, a random number in the Ni block and a Diffie–Hellman value in the KEi block. The initiator's new SPI index is provided in the SA block (Figure 4.13).

The responder confirms the offer in the SA block. It transmits a Diffie–Hellman value in the KEr block. The responder's new SPI index is provided in the SA block (Figure 4.13).

Figure 4.13. *CREATE_CHILD_SA exchange renewal of IKE SA key*

The new SKEYSEED key is calculated from the old SK_d key, the D-H secret key and the random numbers Ni and Nr.

$$SKEYSEED = PRF (old \ SK\_d, D\text{-}H \ key \mid Ni \mid Nr)$$

To renew the ESP/AH SA key, the messages transmitted by the initiator and the responder are similar to the ones used in the creation of the SA. The initiator's request contains an N block (REKEY_SA) containing the SPI index value of the new SA (Figure 4.14).

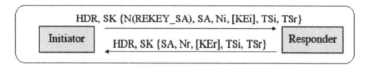

Figure 4.14. *CREATE_CHILD_SA key renewal of ESP / AH SA key*

SSL, TLS and DTLS Protocols

5.1. Introduction

Secure Sockets Layer (SSL) and Transport Layer Security (TLS) protocols are used to create a secure session between a host, which initializes the session, and a security gateway acting as a server, localized in the protected Local Area Network (LAN).

SSL/TLS protocols provide an alternative solution to the Internet Protocol Security (IPSec) mechanism described in Chapter 4.

Secure data transport is implemented between a client (the host) initializing the session and a server (the security gateway), localized in the LAN network protected by SSL/TLS protocols.

The TLS protocol is standardized by the Internet Engineering Task Force (IETF). It succeeds the SSL protocol developed by the Netscape, the original purpose of which was to secure exchanges between a navigator and a web server.

SSL/TLS protocols correspond to a *Record* header inserted between the transport layer and the messages of the application layer, or the following SSL/TLS messages (Figure 5.1):

– the *change_cipher_spec* message indicates a modification of security parameters;

– the *alert* message indicates an error in communication between the host and the security gateway;

– the *handshake* messages negotiate the security parameters between the host and the security gateway.

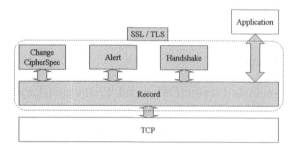

Figure 5.1. *SSL / TLS protocol architecture*

SSL/TLS protocols impose a reliable transport protocol (for example, Transmission Control Protocol (TCP)), which enables the exchange of messages without errors or risks of desequencing.

SSL/TLS protocols delimit messages determining the start and end of each message. Message delimitation is obtained via an indication of length contained in the message. This also enables the transport of multiple SSL/TLS messages in a TCP segment.

The Datagram TLS (DTLS) protocol reuses the main functionalities of SSL/TLS protocols. The modifications it contributes arise from the use of User Datagram Protocol (UDP), Datagram Congestion Control Protocol (DCCP), Stream Control Transmission Protocol (SCTP) and Secure Real-time Transport Protocol (SRTP).

5.2. SSL/TLS protocols

5.2.1. *Record header*

The *Record* header encapsulates SSL/TLS messages or messages originating from the application layer. These messages can be completely encrypted, checked and authenticated due to a seal contained in the Message Authentication Code (MAC) field.

The structure of the *Record* header is shown in Figure 5.2.

The *Protocol* field, coded on 1 byte, indicates the type of message encapsulated:

– value = 20 for the *change_cipher_spec* message;

– value = 21 for the *alert* message;

– value = 22 for the *handshake* messages;

– value = 23 for the message generated by the application.

The *Version* field, coded on 2 bytes, indicates the two values of the protocol version. The latest version of the SSL protocol has a value of {3,0}. The first version of the TLS 1.0 protocol has a value of {3,1}. The current version of the TLS 1.2 protocol has a value of {3,3}.

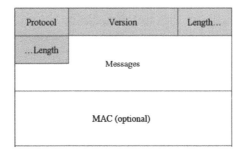

Figure 5.2. *Record header of SSL / TLS protocol*

The *Length* field, coded on 2 bytes, indicates the size of the data encapsulated. This size cannot exceed 2^{14} bytes.

5.2.2. *Change_cipher_spec message*

The *change_cipher_spec* message is simple and defines a single type of message indicating that a new key and a new set of algorithms are being used for message protection. The *change_cipher_spec* message includes a field coded on 1 byte with a value of 1 (Figure 5.3).

Figure 5.3. *Change_cipher_spec message*

5.2.3. *Alert message*

The *alert* message is coded on 2 bytes and contains the *Level* and *Description* fields, with each field coded on 1 byte (Figure 5.4).

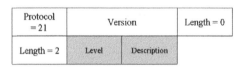

Figure 5.4. *Alert message*

The *Level* field indicates the level of severity of the alarm. Two values are defined:

– a value of 1 indicates a minor alarm. The two parties can continue the current session, but they are not required to establish a new transaction;

– a value of 2 indicates a major alarm. The two parties must end the current session.

The *Description* field indicates the type of alarm encountered. Table 5.1 gives a list of the alarms defined for the SSL protocol.

Name	Value	Description
close_notify	0	This minor alarm tells the receiver not to send any more messages for the current transaction.
unexpected_message	10	This major alarm indicates that an inappropriate message has been received.
bad_record_mac	20	This minor alarm indicates the reception of a *Record* message with an incorrect MAC seal value.
decompression_failure	30	This major alarm indicates that data decompression has failed.
handshake_failure	40	This major alarm indicates that the *handshake* procedure has failed.
no_certificate	41	This minor alarm is transmitted by the client to indicate that it does not possess the certificate required by the server.
bad_certificate	42	This minor alarm indicates that the certificate received is corrupted.
unsupported_certificate	43	This minor alarm indicates that the certificate received is not supported.
certificate_revoked	44	This minor alarm indicates that the certificate received has been revoked.
certificate_expired	45	This minor alarm indicates that the certificate is no longer valid.
certificate_unknown	46	This minor alarm indicates that a problem has occurred with the certificate.
illegal_parameter	47	This major alarm indicates that a field has a prohibited value during the *handshake* procedure.

Table 5.1. *Alarms defined by the SSL protocol*

The TLS protocol has removed the *no_certificate* alarm and added the list of alarms shown in Table 5.2.

Name	Value	Description
record_overflow	22	This major alarm indicates that the size of the *Record* message is too large.
unknown_ca	48	This major alarm indicates that the certification authority is unknown.
access_denied	49	This major alarm indicates that access is refused even though a valid certificate has been received.
decode_error	50	This major alarm indicates that the message cannot be interpreted.
decrypt_error	51	This major alarm indicates that the authentication operation has failed.
protocol_version	70	This major alarm indicates that the protocol version is not supported.
insufficient_security	71	This major alarm indicates that the *handshake* procedure has failed because the client algorithms do not provide adequate security.
internal_error	80	This major alarm indicates that an internal error has interrupted the *handshake* procedure.
user_canceled	90	This minor alarm indicates that the transaction has been terminated.
no_renegotiation	100	This minor alarm is transmitted to indicate rejection of a new negotiation.
unsupported_extension	110	This major alarm is transmitted by the client if the server proposal in the *server_hello* message is different than the proposals initiated by the client.

Table 5.2. *Alarms added by the TLS protocol*

5.2.4. *Handshake messages*

The *handshake* message fulfills the following functions:

– the parties (client and server) exchange *hello* messages to define algorithms and transmit random numbers;

– the parties exchange cryptographic parameters enabling them to build the PreMaster secret;

– the parties exchange certificates and cryptographic parameters enabling their mutual authentication;

– the parties generate the Master key from the PreMaster secret;

– the parties confirm that they have calculated the same security parameters.

The *handshake* messages start with the *Type* field, coded on 1 byte and indicating the type of message, and the *Message Length* field, coded on 3 bytes and indicating the message size excluding the *Type* and *Message Length* fields.

Table 5.3 shows the value of the *Type* field for the different *handshake* messages.

Type field	Message
1	*client_hello*
2	*server_hello*
11	*server_certificate*
12	*server_key_exchange*
13	*certificate_request*
14	*server_hello_done*
15	*certificate_verify*
16	*client_key_exchange*
20	*finished*

Table 5.3. *Different handshake messages*

The *client_hello* and *server_hello* messages insert the protocol version number in the *Protocol Version* field. Though this field appears in the header of the *Record* protocol, it enables the *Record* protocol and the *handshake* messages to evolve independent of one another.

The *handshake* procedure is described in section 2.5.2 concerning the EAP-TLS procedure.

5.2.4.1. *Hello_request message*

The *hello_request* message is transmitted by the server to initialize the *handshake* procedure. The *hello_request* message is a special message containing no fields.

5.2.4.2. *Client_hello message*

The *client_hello* message is used by the client to start the transaction. This message is transmitted upon reception of the *hello_request* message or at the client's initiative in order to renegotiate security parameters.

The structure of the *client_hello* message is shown in Figure 5.5.

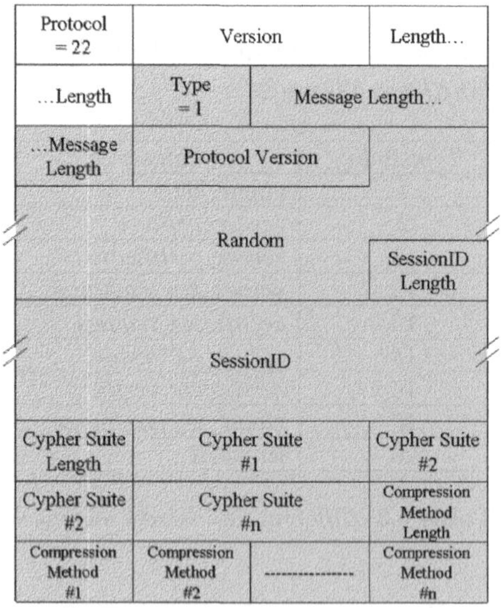

Protocol = 22	Version		Length...
...Length	Type = 1	Message Length...	
...Message Length	Protocol Version		
Random			
		SessionID Length	
SessionID			
Cypher Suite Length	Cypher Suite #1		Cypher Suite #2
Cypher Suite #2	Cypher Suite #n		Compression Method Length
Compression Method #1	Compression Method #2	----------------	Compression Method #n

Figure 5.5. *Client_hello message*

The *Random* field, coded on 32 bytes, contains a random number. The first four bytes contain the date and time down to the second, which makes it possible to avoid duplication of this field's value.

The *SessionID* Length field, coded on 1 byte, indicates the size of the SessionID field.

The *SessionID* field contains the value of the identifier of the session the client wishes to establish. This field is empty if the client wishes to negotiate new security parameters. The client fills this field with the value communicated by the server if it wishes to recover a previous session.

The *Cipher Suite Length* field, coded on 1 byte, indicates the total size of the *Cipher Suite* fields containing proposed algorithmic series. Each *Cipher Suite* field is coded on 2 bytes.

Table 5.4 shows the algorithms proposed by SSL and TLS protocols.

Security functions	SSL	TLS
Key establishment	RSA DH-RSA DH-DSS DHE-RSA DHE-DSS DH-Anon Fortezza-KEA	RSA DH-RSA DH-DSS DHE-RSA DHE-DSS DH-Anon
Encryption	IDEA-CBC RC4-128 3DES-EDE-CBC Fortezza-CBC	IDEA-CBC RC4-128 3DES-EDE-CBC Kerberos AES
Signature	RSA DSA	RSA DSA EC
Hashing	MD5 SHA-1	MD5 SHA-1

Table 5.4. *Algorithmic suites*

The *Compression Method Length* field, coded on 1 byte, indicates the total size of the *Compression Method* fields containing proposed compression methods. Each

Compression Method field is coded on 1 byte. No compression method is currently defined if:

– the value of the *Compression Method Length* field is 1;

– the value of the *Compression Method* field is 0.

The *client_hello* message can request extended functionalities following the list of compression methods from the server. If these functionalities are not available from the server, the *handshake* procedure must be abandoned.

5.2.4.3. *Hello_server message*

The *hello_server* message is the server's response to the *hello_client* message received. The *hello_server* message has a structure similar to the *hello_client* message. The server chooses an algorithm suite and a compression method from the lists proposed by the client.

The server can indicate a value in the *SessionID* field. In this case, the client is authorized to recover this session subsequently. In the opposite case, this field has a value of zero.

5.2.4.4. *Certificate message*

The *certificate* message contains a hierarchical list of certificates:

– the first certificate is the transmitter's;

– the next certificate is the certificate of the authority certifying the previous certificate;

– the third certificate, if there is one, is the authority's certification.

The structure of the *certificate* message is shown in Figure 5.6.

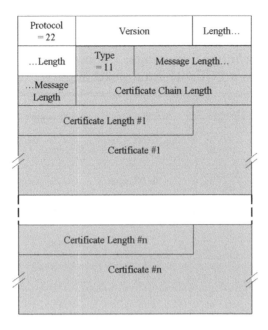

Figure 5.6. *Certificate message*

The value of the *Certificate Chain Length* field, coded on 3 bytes, is equal to the value of the *Message Length* field reduced by 3 bytes.

The *Certificate Length* field, coded on 3 bytes, indicates the size of the certificate.

If the client has produced a list of message signature algorithms in the *client_hello* message extension, the server must sign the list of certificates provided.

5.2.4.5. *Server_key_exchange message*

The *server_key_exchange* message transports cryptographic information from the server to the client, which will enable the client to generate and transmit the PreMaster secret.

Cryptographic information depends on the way in which the key is generated. The structure of the *server_key_exchange* message, shown in Figure 5.7, corresponds to a Diffie–Hellman exchange.

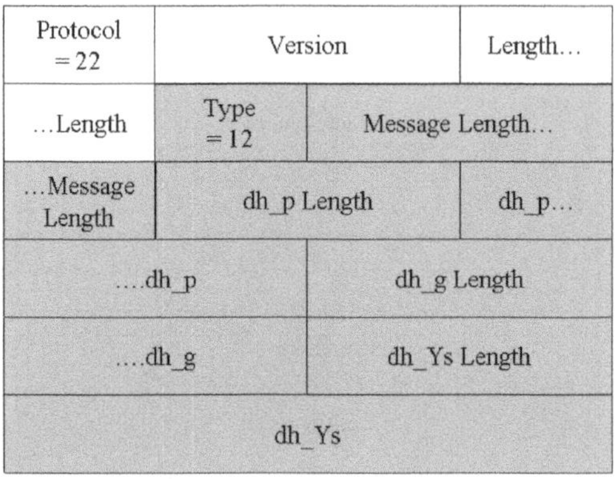

Protocol = 22	Version		Length…
…Length	Type = 12	Message Length…	
…Message Length	dh_p Length		dh_p…
....dh_p		dh_g Length	
....dh_g		dh_Ys Length	
dh_Ys			

Figure 5.7. *Server_key_exchange message*

The *dh_p* field contains the value of the prime number *p*. The length of this field is indicated by the *dh_p Length* field.

The *dh_g* field contains the value of the generator *g*. The length of this field is indicated by the *dh_g Length* field.

The *dh_Ys* field contains the public Diffie–Hellman value of the server. The length of this field is indicated by the *dh_Ys Length* field.

The *server_key_exchange* message can also include a signature after the *dh_Ys* field. The signature's parameters depend on the algorithm indicated in the server's certificate.

5.2.4.6. *Certificate_request message*

The *certificate_request* message is transmitted by the server to request the client's certificate. The *certificate_request* message also indicates the appropriate certification authorities.

The structure of the *certificate_request* message is shown in Figure 5.8.

The *Certificate Type Length* field, coded on 1 byte, indicates the total size of the group of *Certificate Type* fields.

The *Certificate Type* field, coded on 1 byte, provides the type of certificate:

– value = 1: RSA, for key exchange signature;

– value = 2: DSA, for signature only;

– value = 3: RSA, for signature with Diffie–Hellman fixed-key exchange;

– value = 4: DSA, for signature with Diffie–Hellman fixed-key exchange;

– value = 5: RSA, for signature with Diffie–Hellman temporary-key exchange;

– value = 6: DSA, for signature with Diffie–Hellman temporary-key exchange;

– value = 20: Fortezza/DMS, for signature and key exchange.

The *Certificate Authorities Length* field, coded on 2 bytes, contains the total size of the group of *Certificate Authorities* fields. Each *Certificate Authorities* field is also preceded by the *Certificate Authorities Length* field indicating its size.

The *Certificate Authorities* field contains *Distinguished Name* information concerning the certification authority contained in the certificate.

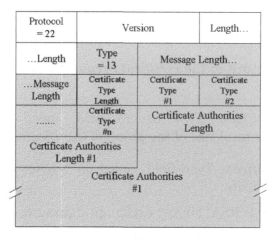

Figure 5.8. *Certificate_request message*

5.2.4.7. *Server_hello_done message*

The *server_hello_done* message is transmitted by the server to indicate the end of the *server_hello* message and the associated *certificate*, *server_key_exchange* and *certificate_request* messages.

5.2.4.8. *Client_key_exchange message*

The *client_key_exchange* message is transmitted by the client just after the *certificate* message containing the client's certificate. The *client_key_exchange* message is used to establish the Premaster secret.

The structure of the *client_key_exchange* message is shown in Figure 5.9 if this secret is established using Diffie–Hellman exchanges.

The *dh_Yc* field contains the client's public Diffie–Hellman value. The length of this field is indicated by the *dh_Yc Length* field. This field appears when the value is temporary. If the value is fixed, it appears in the client's certificate, and in this case the field is empty.

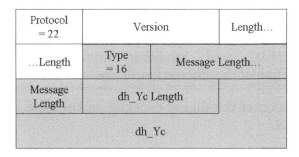

Protocol = 22	Version		Length...
...Length	Type = 16	Message Length...	
Message Length	dh_Yc Length		
dh_Yc			

Figure 5.9. *Client_key_exchange message*

5.2.4.9. *Certificate_verify message*

The *certificate_verify* message is transmitted by the client to prove that it holds the private key corresponding to the public key of its certificate. The *certificate_verify* message contains the signature of a digest calculated from all of the *handshake* messages exchanged since the *client_hello* message and from the Master secret derived from the PreMaster secret.

Calculation of the digest depends on the type of protocol: SSL or TLS.

5.2.4.10. *Finished message*

After the *certificate_verify* message, the client transmits the *change_cipher_spec* message to indicate that the next message, the *finished* message, uses the algorithm suite and key negotiated for the encryption and integrity control functions.

The *finished* message enables the client to terminate the negotiation. It contains the following encrypted information:

– the digest calculated from all of the *handshake* messages exchanged since the *client_hello* message, a sequence of characters depending on the role of the party (client or server), and the Master secret;

– the MAC seal enabling integrity control and authentication of the *finished* message.

Calculation of the digest depends on the type of protocol: SSL or TLS.

5.2.5. *Cryptographic information*

5.2.5.1. *Key generation*

The Master key is derived from the PreMaster key, the implementation of which uses one of the following operations:

– the client and server exchange public Diffie–Hellman values. This mechanism is described in section 5.2.4 concerning *handshake* messages.

– the client generates the PreMaster secret and transmits it to the server, encrypting it using the server's public key (RSA or Fortezza/DMS mechanisms).

For the SSL protocol, Master key is calculated using message digest 5 (MD5) and Secure Hash Algorithm (SHA) for the calculation of digests.

```
master secret =
MD5(premaster secret || SHA(ASCII character 'A' ||
premaster secret || client_hello.random ||
server_hello.random)) ||
MD5(premaster secret || SHA(ASCII characters 'BB' ||
premaster secret || client_hello.random ||
server_hello.random)) ||
MD5(premaster secret || SHA(ASCII characters 'CCC' ||
premaster secret || client_hello.random ||
server_hello.random))
```

For the TLS protocol, the Master key is calculated using the Hashed message authentication code (HMAC) operation for the calculation of digests.

The next formula is the one defined for the TLS 1.2 protocol.

It differs from the formula used for previous versions, which used a combination of two SHA and MD5 algorithms.

The number of iterations depends on the required size of cryptographic information to be produced.

```
master secret =
HMAC-SHA-256(premaster secret, A(1) | | label | | seed) | |
HMAC-SHA-256 (secret, A(2) | | label | | seed) | |
HMAC-SHA-256 (secret, A(3) + label | | seed) + ...
with
A(0) = label | | seed
A(i) = HMAC(secret, A(i-1))
```

The client and the server then derive the Master secret to generate the following keys (key materials):

– *client write MAC secret:* this secret is used by the client to calculate the MAC seal associated with a message;

– *server write MAC secret:* this secret is used by the server to calculate the MAC seal associated with a message;

– *client write encryption key*: this key is used by the client to encrypt the message and associated MAC seal;

– *server write encryption key*: this key is used by the server to encrypt the message and associated MAC seal;

– *client write IV*: this structure is used by the client as an initialization vector for the encryption algorithm;

– *server write IV*: this structure is used by the server as an initialization vector for the encryption algorithm.

The calculation mechanism is identical to the mechanism used to derive the PreMaster key.

5.2.5.2. *Integrity checking*

For the SSL protocol, the MAC seal is calculated by applying the MD5 algorithm to the following concatenated data:

– the *write MAC* secret, derived from the Master secret;

– byte 36 in hexadecimal notation, repeated 48 times;

– the number of messages exchanged (*Sequence Number*). The counter value is set at zero for each *change_cipher_spec* message, and then increased by 1 unit for each message;

– the *Protocol* and *Length* fields of the *Record* header;

– the message being subjected to integrity checking.

For the TLS protocol, the MAC seal calculation is shown in Figure 5.10.

5.3. DTLS protocol

5.3.1. *Adaptation to UDP transport*

5.3.1.1. *Record header*

The TLS protocol implicitly maintains a message sequence number that is never transmitted. However, this number is used for calculating the MAC seal. The TCP protocol ensures the delivery of segments in sequence. Since this function does not exist for the UDP protocol, message numbering has been introduced by the DTLS protocol.

The structure of the *Record* header is shown in Figure 5.11.

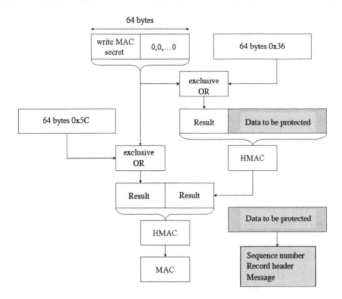

Figure 5.10. *Seal calculation for TLS protocol*

Protocol	Version	Epoch...
...Epoch	Sequence Number....	
Sequence Number....		Length...
...Length		
	Messages	
	MAC (optional)	

Figure 5.11. *Record header of DTLS protocol*

The *Epoch* field, coded on 2 bytes, is increased by 1 unit for each successful *handshake* procedure.

The *Sequence Number* field, coded on 6 bytes, is increased by 1 unit for each *Record* message.

5.3.1.2. *Handshake messages*

As for the TLS protocol, the *handshake* procedure starts when the *client_hello* message is sent. Since there is no connection establishment procedure by the transport protocol, there is a risk of denial of service type attack. An attacker can send *client_hello* messages and thus open multiple sessions with the server.

To avoid this problem, the DTLS protocol introduces a new *hello_verify_request* message containing a reference (*cookie*), transmitted by the server in response to the *client_hello* message.

The client must repeat the *client_hello* message, integrating the reference (*cookie*) received after the *SessionID* field. The server can then verify that the client has not usurped its address.

The TCP protocol segments messages that are too long in order to adapt itself to the authorized size. Since this function does not exist for the UDP protocol, the DTLS protocol fragments *handshake* messages so that this operation will not take place in the Internet Protocol (IP) layer.

The DTLS protocol must determine the Maximum Transmission Unit (MTU) size of the IP packet on the route between the client and the server in order to deduce the maximum size of the fragment from it.

The TCP protocol ensures retransmission in case of segment loss. Since this function does not exist for the UDP protocol, the DTLS protocol must arm a timer and retransmit the message when the timer has expired.

The structure of the *handshake* message header is shown in Figure 5.12.

The *Message Sequence Number* field, coded on 2 bytes, contains the number of the *handshake* messages. The value of this field is constant for each message fragment.

The *Fragment Offset* field, coded on 3 bytes, indicates the total number of bytes in all previous fragments. The value of this field is zero if the message is not fragmented.

The *Fragment Length* field, coded on 3 bytes, indicates the size of the fragment. The value of this field is equal to the value of the *Message Length* field if the message is not fragmented.

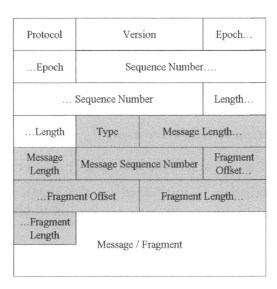

Figure 5.12. *Handshake message header in DTLS protocol*

5.3.2. *Adaptation to DCCP transport*

The DCCP protocol has the following characteristics:

– as for the TCP protocol, the transmission of data is contingent on the opening of a session between the client and the server;

– as for the UDP protocol, the data transmitted is not subject to retransmission in the event of loss;

– the data transmitted is the subject of an acknowledgment, the purpose of which is not the retransmission of missing data but the provision of information on data delivered to the destination;

– the protocol's functionalities are the subject of a negotiation;

– the congestion checking mechanism is chosen by the application layer.

As for TLS with regard to TCP, the DCCP connection is made before the *handshake* procedure of the DTLS protocol. However, it is possible to transmit data during a DCCP connection. In this case, the data transmitted is composed of the *handshake* procedure messages.

5.3.3. *Adaption to SCTP transport*

The SCTP protocol has the following characteristics:

– an association is initialized by the client. During the initialization phase, a reference (*cookie*) is implemented for protection against attacks.

– as for the TCP protocol, data is delivered to the application without error and in sequence. If the transfer executed by the TCP protocol is byte-oriented, the transfer executed by the SCTP protocol is block oriented.

– as for the TCP protocol, the source output is regulated either by the receiver or by the *Slow-Start* and *Congestion Avoidance* congestion control mechanism;

– fragmentation is ensured by the transmitter to adapt to the minimum size encountered in the network;

– messages from multiple applications can be multiplexed in a single SCTP connection;

– multiple interfaces per host can participate in association (*multi-homing*).

Though the SCTP protocol functions like the TCP protocol, it uses the DTLS protocol rather than the TLS protocol. This is because the SCTP protocol has specific functionalities that are not compatible with the TLS protocol:

– the SCTP protocol can transport messages that are not in sequence;

– the SCTP protocol can transport messages without an acknowledgment;

– the SCTP protocol can transport one-way messages.

Some functions executed by the DTLS protocol are removed when they are already executed by the SCTP protocol:

– retransmission in case of loss;

– finding of the maximum MTU value.

5.3.4. *Adaption to SRTP transport*

The Real-time Transport Protocol (RTP) is the protocol used by real-time applications (voice and video). It is a complementary transport protocol to the associated UDP protocol. It provides the following functions:

– identification of type of multimedia information;

– numbering of RTP segments;

– time-stamping of RTP segments.

The RTP Control Protocol (RTCP) is the control protocol associated with the RTP protocol. RTCP messages provide the following information:

– service quality parameter values such as loss rate, delay and jitter of RTP segments;

– transmitter identifier, clock for synchronizing two data sources (voice and video), cumulative segment counter and number of bytes sent.

The SRTP protocol encrypts the data encapsulated by the RTP header and RTCP messages, as well as checking the integrity of RTP segments or RTCP messages.

The DTLS protocol provides a frame for the SRTP protocol for security parameter negotiation. However, the encryption and integrity-checking principles defined for the SRTP protocol are maintained, meaning that RTP or RTCP flows are not encapsulated by the *Record* header.

The *handshake* procedure is part of the Session Information Protocol (SIP) procedure used to establish a session (for example, a telephone call). When the RTP flow (containing the voice, for example) has been negotiated via the Session Description Protocol (SDP), the SIP party initializes the sending of the first *client_echo* message of the *handshake* procedure.

The *client_hello* message contains the security parameters proposed by the client. These parameters, specific to the RTP flow, are defined in a specific extension called *use_srtp*. Several profiles are available, specifying the encryption and integrity checking algorithms.

The server's *hello_server* response contains the security parameter finally chosen by the server in the *use_srtp* extension.

6

Network Management

6.1. SNMPv3 management

6.1.1. *Introduction*

Simple Network Management Protocol (SNMP) describes the management architecture of the most frequently used network equipment.

Each item of network equipment includes an entity called the agent, which enables it to access a Management Information Base (MIB) containing objects describing the equipment. These objects are inventoried in a naming tree.

All data stored in the MIB correspond to a variable associated with an object. This variable is described in the language Abstract Syntax Notation (ASN.1). The Structure of Management Information (SMI) defines the representation of each element of information in the MIB.

The equipment management station includes an entity called the manager, which interacts with the agent via the intermediation of the SNMP protocol.

The SNMPv1 protocol is composed of requests generated by the manager, responses from the agent and alert messages from the agent:

– the *GetRequest* message is used to search for a variable in an object in the MIB;

– the *GetNextRequest* message is used to search for the next variable;

– the *SetRequest* message is used to change the value of a variable of an object in the MIB;

– upon reception of a request, the agent always responds with a *GetResponse* message;

– when an unexpected event occurs with the equipment, the agent informs the manager by sending a *Trap* message.

The security of this version is based solely on the *community* field of the SNMPv1 header. The SNMPv1 message is neither integrity-checked nor encrypted.

SNMPv2 architecture attempted to correct a number of flaws in the SNMPv1 architecture:

– the table manipulation was improved due to the SMIv2 structure;

– the protocol was made more efficient due to mass transfer by the *GetBulkRequest* message;

– the introduction of notifications between managers using the *Inform* message.

However, the section pertaining to security defined in SNMPv2p (p for *party*) was not retained. Instead, the security mechanism from the first version was preserved, hence the term SNMPv2c (c for *community*) to designate the second version.

Following SNMPv2p, two rival standards emerged: SNMPv2u and SNMPv2*. Security was eventually defined in the SNMPv3 architecture, which combined these two standards.

6.1.2. *SNMPv3 architecture*

SNMPv3 consists of a group of distributed entities. Each entity acts as a manager or an agent, and each entity is a collection of modules interacting with each other. Modular architecture makes it possible for each module to be developed separately.

Each entity includes an SNMP engine that executes functions pertaining to the transmission of messages, implements security mechanisms and controls MIB access. These functions are seen as services rendered to the different SNMPv3 application modules.

SNMPv3 architecture is shown in Figure 6.1 for the manager and Figure 6.2 for the agent.

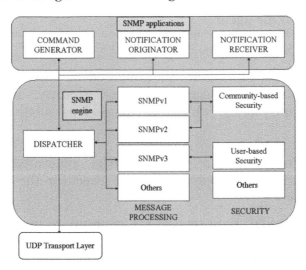

Figure 6.1. *SNMPv3 architecture, manager side*

6.1.2.1. *SNMPv3 applications*

SNMPv3 applications generate and respond to received messages as well as generate and receive notifications. Five types of applications are defined:

– COMMAND GENERATOR: this application generates the *GetRequest*, *GetNextRequest*, *GetBulkRequest* and *SetRequest* messages. It is localized in the manager;

– COMMAND RESPONDER: this application responds to requests received. It is localized in the agent accessing the MIB;

– NOTIFICATION ORIGINATOR: this application generates *Trap* or *Inform* notifications pertaining to events, such as alarms. It is localized in the agent for *Trap* messages accessing the MIB, and in the manager for the *Inform* message;

– NOTIFICATION RECEIVER: this application receives *Trap* or *Inform* notifications and is localized in the manager;

– PROXY FORWARDER: this application is used to transfer messages between two SNMPv3 entities and is localized in the agent.

6.1.2.2. *SNMPv3 engine*

The SNMPv3 engine is composed of the *Dispatcher* module and the *Message Processing*, *Security*, and *Access Control* subsystems, which are themselves subdivided into modules.

For Protocol Data Units (PDUs) generated by the application, the *Dispatcher* module executes the following operations:

– directs PDU data toward the appropriate module of the *Message Processing* subsystem depending on the SNMP message version. After processing, the module of the *Message*

Processing subsystem resends the SNMP message to the *Dispatcher* module;

– transmits the SNMP message received to the user datagram protocol (UDP) transport layer.

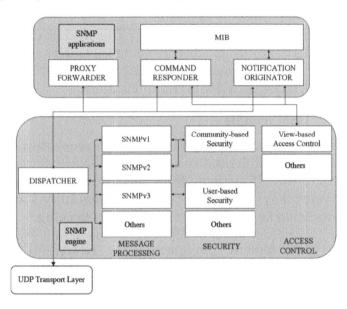

Figure 6.2. *SNMPv3 architecture, agent side*

For SNMP messages received from the transport layer, the *Dispatcher* module executes the opposite operations.

The *Message Processing* subsystem is composed of SNMPv1, SNMPv2 and SNMPv3 modules. For data generated by the application, the SNMP module adds the appropriate header to the PDU data units. For data generated by the transport layer, the SNMP module removes the header from the SNMP messages.

The *Security* subsystem is composed of *User-based Security* and *Community-based Security* modules. The

User-based Security module provides security, integrity-checking and confidentiality services. This module is used by the SNMPv3 module. The *Community-based Security* module is used by the SNMPv1 and SNMPv2c modules.

The *Access Control* subsystem contains only the View-Based Access Control Model (VACM). This module determines whether MIB access is authorized or not. This module is requested by the agent's COMMAND RESPONDER and NOTIFICATION ORIGINATOR applications.

6.1.2.2.1. USM module

The User-based Security Model (USM) provides integrity-checking and confidentiality services for SNMPv3 messages. To execute these functions, the USM module uses two keys; a private key (*privkey*) for confidentiality and an authentication key (*authkey*) for integrity-checking.

The USM module implements the HMAC-MD5-96 or HMAC-SHA-96 mechanisms for integrity control, and the Cipher Block Chaining (CBC) mode of the Data Encryption Standard (DES) algorithm for confidentiality.

Each agent has a unique key for each authorized user. If multiple users can access an agent, the agent has an authentication and encryption key for each user.

If a user accesses multiple agents, it must use different keys. A derivation mechanism is used to generate these keys from a unique password, which means the user does not have to remember a large number of keys:

– a first derivation from the password produces the user key;

– a second derivation from the user key and the agent's engine identifier produce keys used for integrity-checking and confidentiality.

6.1.2.2.2. VACM module

An application invokes the VACM module via the *isAccessAllowed* primitive with the *securityModel*, *securityName*, *securityLevel*, *viewType*, *contextName* and *variableName* parameters.

The *securityModel* and *securityLevel* parameters define the protection mode to be applied to a user's data. A combination of the *securityModel* and *securityName* parameters determines a group (*groupName*) listed in the *securityToGroupTable* table. A couple (*securityName* and *securityModel*) can belong to only one group. A user (*securityName*) can belong to several groups with different security models.

The *contextName* parameter allows access to the *vacmContextTable* context table.

The four elements (group, context name, security level and security model) and type of operation (*Read/Write/Notify*) described in the *viewType* parameter select a view (*viewName*) from the *vacmAccessTable* table, which is an index of the *VacmViewTreeFamilyTable* table.

The *variableName* parameter is an object identifier that identifies a type of object and an instance, for which its presence is checked in the *VacmViewTreeFamilyTable* table.

The VACM procedure is shown in Figure 6.3.

6.1.2.3. *Operation procedures*

Services between modules of the SNMP entity are defined in terms of primitives and parameters. A primitive specifies the function to be carried out. Parameters enable data transfer and control information.

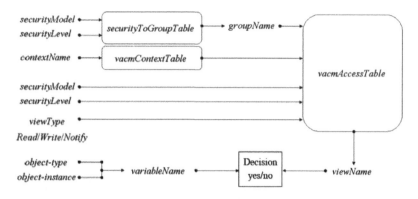

Figure 6.3. *VACM procedure*

The process of emission of the SNMP request is shown in Figure 6.4.

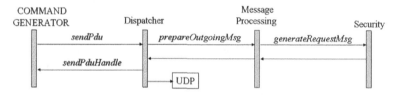

Figure 6.4. *Emission of SNMP request*

The COMMAND GENERATOR application uses the *sendPdu* primitive to supply the *Dispatcher* module with PDU data and control data as the destination and security level to be applied.

The *Dispatcher* module invokes the *Message Processing* subsystem (*prepareOutgoingMsg* primitive), which then requests the *Security* subsystem (*generateRequestMsg* primitive).

The *Dispatcher* module transfers the SNMP message received from the *Message Processing* subsystem to the UDP transport layer and returns a *sendPduHandle* identifier to the application, which is used by the application to correlate requests and responses.

The COMMAND RESPONDER application uses four primitives to communicate with the *Dispatcher* module:

– *registerContextEngineID*: this primitive enables the application to register the context identifier with the *Dispatcher* module;

– *unregisterContextEngineID*: this primitive executes the opposite operation;

– *processPdu*: this primitive is used by the *Dispatcher* module to deliver PDU data from the request to the application;

– *returnResponsePdu*: when the request has been processed, the application uses this primitive to respond to the *Dispatcher* module.

The process of reception of an SNMP request is shown in Figure 6.5.

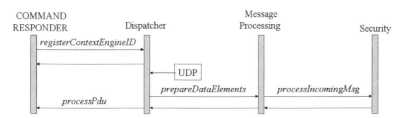

Figure 6.5. *Reception of SNMP request*

The *prepareDataElements* primitive is used by the *Dispatcher* module to transmit the SNMP message received to the *Message Processing* subsystem in order to recover PDU data.

The *processIncomingMsg* primitive is used by the *Security* module to receive the SNMP message in order to check integrity and to decrypt it.

The process of emission of the SNMP response is shown in Figure 6.6.

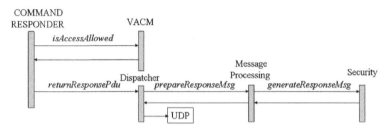

Figure 6.6. *Emission of SNMP response*

After examining the content of the SNMP request, the COMMAND RESPONDER application checks the VACM module and that access is authorized. The COMMAND RESPONDER application uses the *isAccessAllowed* primitive to solicit the VACM module.

If access is authorized, the application prepares the PDU data and transmits it to the *Dispatcher* module in the *returnResponsePdu* primitive.

The SNMP message is prepared by soliciting the *Message Processing* subsystem via the *prepareResponseMsg* primitive, and then the *Security* subsystem by the *generateResponseMsg* primitive.

The process of reception of SNMP response is shown in Figure 6.7.

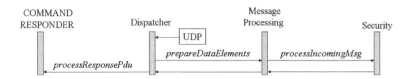

Figure 6.7. *Reception of SNMP response*

On reception of the SNMP message, the same primitives (*prepareDataElements* and *processIncomingMsg*) are used to recover the PDU data transmitted by the *Dispatcher* module to the COMMAND GENERATOR application with the *processResponsePdu* primitive.

6.1.3. *SNMPv3 message structure*

The SNMPv3 message structure is shown in Figure 6.8.

An SNMPv3 message contains a PDU structure identical to the one used for SNMPv1 or SNMPv2 messages. This structure defines the PDU type (request, response and notification) and the variable types and values.

The PDU structure is encapsulated by a header specific to the SNMPv3 protocol and contains the following parts:

– the part concerning processing by the *Message Processing* module;

– the part concerning processing by the *User Security* module;

– the part concerning context identification.

The *msgVersion* field, coded on 4 bytes, contains the protocol version. It has a value of 3.

The *msgID* field, coded on 4 bytes, contains an identifier used to correlate requests and responses. This field is also used by the *Message Processing* module to coordinate the processing of the message by different modules.

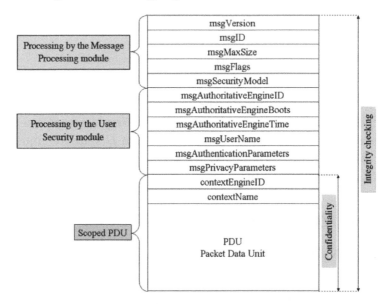

Figure 6.8. *SNMPv3 message structure*

The *msgMaxSize* field, coded on 4 bytes, indicates the maximum size of the message supported by the generator. It has a value between 484 and 2^{31}-1.

The *msgFlags* field, coded on 1 byte, contains the following flags:

– *reportable-Flag*: this bit, positioned at one, indicates that the destination must respond to the PDU data received. *GetRequest, GetBulkRequest, SetRequest* and *Inform* data position this bit at one;

– *privFlag*: this bit, positioned at one, indicates that *Scoped*PDU data are encrypted;

– *authFlag*: this bit, positioned at one, indicates that the message has been integrity-checked and authenticated. The combination of *privFlag* = one and *authFlag* = zero is forbidden.

The *msgSecurityModel* field, coded on 4 bytes, indicates which security module has been used. It has a value of 3 for the SNMPv3 protocol.

The *msgAuthoritativeEngineID* field contains the destination identifier for *GetRequest*, *GetBulkRequest*, *SetRequest* or *Inform* PDU data, or for the source of *Trap* or *Response* PDU data.

Structure:
```
04 <length> <msgAuthoritativeEngineID>
```
Example:
```
04 09 80000002   IBM
      01          IPv4 address
      09840301    9.132.3.1
```

The *msgAuthoritativeEngineBoots* field indicates the number of reinitializations since the initial configuration.

Structure:
```
02 <length> <msgAuthoritativeEngineBoots>
```
Example:
```
02 01 01     1 boot
```

The *msgAuthoritativeEngineTime* field indicates the number of seconds elapsed since the last reinitialization.

Structure:
```
02 <length> < msgAuthoritativeEngineTime>
```
Example:
```
02 02 0101     257 seconds
```

The *msgUserName* field indicates the user name on behalf of which the message was generated.

Structure:

```
02 <length> < msgUserName>
```

Example:

```
02 04 62657274    bert
```

The *msgAuthenticationParameters* field indicates the parameters used for authentication and integrity-checking of the message.

The *msgPrivacyParameters* field contains the initialization vector of the DES CBC encryption algorithm.

The *contextEngineID* field contains the context identifier.

The *contextName* field contains the context name.

6.2. SSH protocol

The Secure Shell (SSH) protocol is used to secure the Transmission Control Protocol (TCP) connection. It includes three components:

– the SSH-TRANS transport protocol is used to authenticate the server and to implement data protection and compression algorithms and keys;

– the SSH-USERAUTH protocol is used to authenticate the client. It relies on the SSH-TRANS protocol;

– the SSH-CONNECT protocol is used to multiplex multiple sessions. It relies on the SSH-TRANS protocol.

6.2.1. *SSH-TRANS protocol*

The SSH-TRANS protocol relies on the TCP protocol. The client uses the number 22 as destination port.

The structure of the SSH-TRANS header is shown in Figure 6.9.

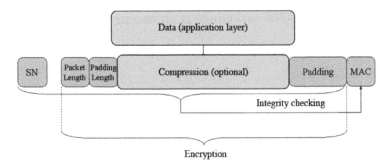

Figure 6.9. *Structure of SSH-TRANS header*

When the SSH connection is established, the client and the server can exchange messages originating from the application layer encapsulated by the SSH-TRANS header, including the following fields:

– *Packet length*: this field codes the size of the data originating from the application layer, which may be compressed, the padding and the *Padding length* field;

– *Padding length*: this field codes the size of the padding;

– *Padding*: this field is used to complete the data originating from the application layer in order to obtain a multiple size of the block being encrypted (*block cipher*) or a multiple of eight octets in the case of encryption by flow (*stream cipher*);

– Message Authentication Code (MAC): this field contains the seal calculated from the SSH-TRANS protocol data and a Sequence Number (SN). The SN field is set at zerofor the first unit of SSH-TRANS data transmitted and increased by one unit for each unit of SSH-TRANS data. This field is not transmitted in the SSH-TRANS header.

Encryption is applied to the *Packet length*, *Padding length*, and *Padding* fields and to data units originating from the application layer.

The SSH-TRANS procedure is shown in Figure 6.10.

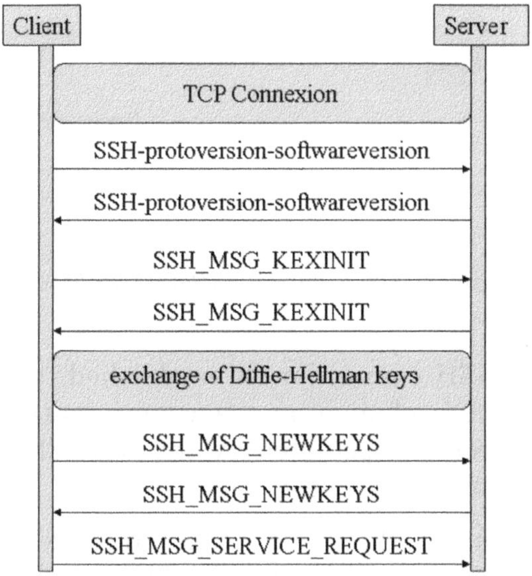

Figure 6.10. *SSH-TRANS procedure*

The first stage consists of the exchange of an identification message between the client and server, encapsulated by the SSH-TRANS header, in the following form:

– *SSH-protoversion-softwareversion SP comments CR LF*, with;

– Space (SP) corresponds to a space;

– Carriage Return (CR) corresponds to the ENTER key;

– Line Feed (LF).

The *comments* field is optional. In the example below, the *comments* field has been omitted:

SSH-2.0-billsSSH_3.6.3q3<CR><LF>

The part before the CR field constitutes the identification of the end point. It is used during the exchange of Diffie–Hellman keys.

The next stage concerns the negotiation of algorithms. Each end point transmits the SSH_MSG_KEXINIT message with a list of algorithms in preferential order. These algorithms concern key exchange, encryption, calculation of the authentication seal and compression. For each category, the server chooses the first algorithm proposed by the client and supported by the server.

The exchange of keys occurs via the following three exchanges:

– the client generates a number x $(1 < x < q)$ and calculates $e = g^x \bmod p$. The client transmits the value e to the server. The values p, g and q are known by the client and server during the negotiation of algorithms;

– the server generates a number y $(0 < y < q)$ and calculates $f = g^y \bmod p$. It calculates the master key $K = e^y \bmod p$, digest H = HASH(V_C || V_S || I_C || I_S || K_S || e || f || K) and signature s of digest H with its private key. The server transmits the values (K_S || f || s) to the client. The digest H, the hash function of which is decided during algorithm negotiation, is calculated using the following parameters:

- V_C = client identification,

- V_S = server identification,

- I_C = SSH_MSG_KEXINIT message from the client,

- I_S = SSH_MSG_KEXINIT message from the server,

- K_S = server's public key.

– the client can verify the server's public key K_S using a certificate. It calculates the master key $K = f^x \bmod p$ and digest H = HASH(V_C || V_S || I_C || I_S || K_S || e || f || K), and verifies the signature s of the digest H.

After keys have been exchanged, the server is authenticated with its private key. The two end points exchange the SSH_MSG_NEWKEYS message to signal the start of the implementation of the master key K generated during the key exchange.

The last stage deals with the service request. The client transmits the SSH_MSG_SERVICE_REQUEST message in order to start the SSH-USERAUTH or SSH-CONNECT procedure.

At the end of the SSH-TRANS procedure, data exchanged between the client and server are encapsulated by the SSH-TRANS header, encrypted and authenticated. Initialization vectors and keys are generated from the master key K, the value of the digest H calculated during the key exchange, a character (A to F) and the *session_id* session identifier (equal to the digest H):

– initialization vector from client to server: HASH(K || H || "A" || session_id);

– initialization vector from server to client: HASH(K || H || "B" || session_id);

– encryption key from client to server: HASH(K || H || "C" || session_id);

– encryption key from server to client: HASH(K || H || "D" || session_id);

– integrity key from client to server: HASH(K || H || "E" || session_id);

– integrity key from server to client: HASH(K || H || "F" || session_id).

6.2.2. *SSH-USERAUTH protocol*

At the end of the SSH-TRANS procedure, the user authentication procedure can begin. It consists of the exchange of the following messages:

– the client transmits the SSH_MSG_USERAUTH_REQUEST message containing its identity and the service (SSH-CONNECT protocol) the client wishes to access;

– the server checks the client's identity:

 - if the server rejects the client's request, it responds with the SSH_MSG_USERAUTH_FAILURE message, with the *partial success* field given a value of FALSE,

 - if the server accepts the client's request, it responds with the SSH_MSG_USERAUTH_FAILURE message, with the *partial success* field given a value of TRUE, and proposes authentication methods;

– the client selects one of the methods proposed and transmits the SSH_MSG_USERAUTH_REQUEST message containing the method chosen. At this stage, exchanges can take place to complete authentication;

– if authentication has succeeded and additional authentication methods are required, the server responds with the SSH_MSG_USERAUTH_FAILURE message, with the *partial success* field given a value of TRUE;

– if authentication has failed, the server responds with the SSH_MSG_USERAUTH_FAILURE message, with the *partial success* field given a value of FALSE;

– if all authentication methods are successful, the server responds with the SSH_MSG_USERAUTH_SUCCESS message, which ends the SSH-USERAUTH procedure.

6.2.3. *SSH-CONNECT protocol*

At the end of the two SSH-TRANS and SSH-USERAUTH phases, the SSH-CONNECT procedure can start. It is based on messages exchanged between the client and server, encapsulated by the SSH-TRANS header.

The SSH-CONNECT protocol makes it possible to multiplex multiple sessions by using separate channels. Each channel can be opened by the client or server and is associated with a number, the value of which can be different for each end point. Channel flow is controlled, and neither end point can transmit before receiving a message indicating the receiver's available memory space.

The SSH-TRANS procedure is composed of three stages (Figure 6.11):

– channel opening;

– transfer of session data;

– channel closing.

The local end point wishing to open a channel transmits the SSH_MSG_CHANNEL_OPEN message, indicating the channel number, initial size of reception window and maximum size of the SSH-CONNECT PDU for traffic received by the initiator of the channel opening.

Four types of channels have been defined:

– *session*: the session refers to the execution of a remote program. The program can be a *shell* (software layer providing an interface for the use of the operating system) or an application such as, for example, file transfer or e-mail;

– *x11*: this type of channel is used when applications operate on a server and are displayed on a computer;

– *direct-tcpip*: this type of channel is used to configure a port number, as the example below indicates:

- the client-side application wishing to open a post office protocol (POP3) session with a messaging server uses port 110 as its destination port,

- SSH is configured to convert the port number 110 into an unattributed number, for example, 9999;

– *forwarded-tcpip*: this type of channel is used to configure the port number, as shown in the example below:

- the client-side application is configured with a source port number, for example, 2222, with the destination port number being 22,

- the server-side application is configured to receive traffic from source 2222.

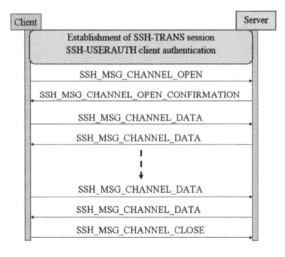

Figure 6.11. *SSH-CONNECT procedure*

If the request is accepted, the remote end point returns the SSH_MSG_CHANNEL_OPEN_CONFIRMATION   message containing the channel number defined by the local end point, the channel number defined by the remote end point, the initial size of the reception window and the maximum size of the SSH-CONNECT PDU for traffic received by the remote end point.

In the opposite case, the remote end point responds with the SSH_MSG_CHANNEL_OPEN_FAILURE message indicating the reason for failure.

When the channel is open, data transfer is executed using the SSH_MSG_CHANNEL_DATA message containing the channel number defined by the destination end point.

The closing of the channel can be activated by either of the end points via the sending of the SSH_MSG_CHANNEL_CLOSE message.

7

MPLS Technology

7.1. MPLS overview

7.1.1. *Network architecture*

In a Multi-Protocol Label Switching (MPLS) network, packets are transferred from a label (or identifier) generally corresponding to a destination Internet Protocol (IP) address. Labels can be generated for a router or for each interface of the router. They have only a local meaning and not an end-to-end meaning. Labels are used to construct a virtual circuit called a Label Switching Path (LSP).

The basic principle of MPLS is label switching. These labels, which are simple integers, are inserted between the headers of levels 2 and 3 and enable switching from these labels with no need to consult the IP header or routing table. This technique is similar to the switching of cells in Asynchronous Transfer Mode (ATM) with Virtual Path Identifier/Virtual Channel Identifier (VPI/VCI), or to switching on a Frame Relay network with the Data Link Connection Identifier (DLCI). MPLS authorizes the creation of label stacks.

A MPLS network is built from two types of routers (Figure 7.1):

– the *Edge* Label Switching Router (LSR) is located on the periphery of the MPLS network. The *Edge* LSR router for the input flow in the network is called the *ingress Edge* LSR, and that for the output flow is called the *egress Edge* LSR;

– the *Core* LSR router is located in the core network.

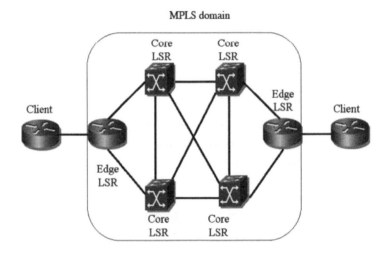

Figure 7.1. *MPLS network architecture*

Edge LSR routers execute IP routing (Figure 7.2). The input LSR router inserts a label (PUSH function), and the output LSR router removes the label (POP function).

At the MPLS network ingress, IP packets are grouped into Forwarding Equivalent Classes (FEC). Packets belonging to the same FEC will follow the same route in the MPLS network. Typically, FEC are linked to IP addresses learned by intradomain routing protocols operating on the MPLS network. There is an LSP for each FEC and LSPs are one directional.

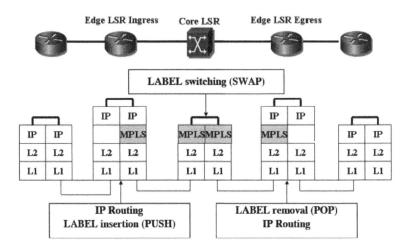

Figure 7.2. *MPLS network protocol architecture*

7.1.2. *LSR router tables*

Information necessary to reach the destination IP network is stored in the Routing Information Base (RIB) routing table. The RIB table is localized in the control plane of the LSR router (processing card). Information stored in the RIB table is transferred to another table, the Forwarding Information Base (FIB) table. The FIB is localized in the user plane of the LSR router (line card). By combining several tables including the RIB table, the FIB table can increase the speed of data transfer (Figure 7.3).

The table Label Information Base (LIB) is informed by the Label Distribution Protocol (LDP). It contains links between the label and the FEC. The LIB table is localized in the control plane of the LSR router. A router port is called *downstream* if this port corresponds to the egress port of the FIB table. The other ports are called *upstream*.

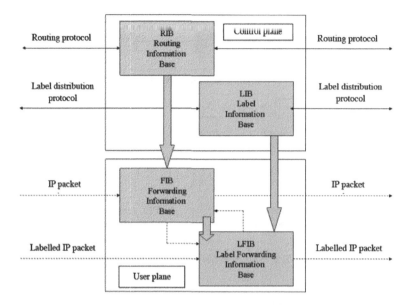

Figure 7.3. *LSR router tables*

The Label Forwarding Information Base (LFIB) table is fed by the FIB and LIB tables. It is used by *Core* LSR routers to execute label switching, by the *ingress Edge* LSR router to insert the label; and by the *egress Edge* LSR router to remove the label. The LFIB table is localized in the user plane of the LSR router.

7.1.3. *PHP function*

The Penultimate Hop Popping (PHP) function consists of removing the MPLS header from between the *egress Edge* LSR router (output router) and the LSR router located upstream (Figure 7.4). If the packets received by the *egress Edge* LSR are labeled, they must first remove the label using the LFIB table, then execute a search in the FIB transfer table to find the output interface.

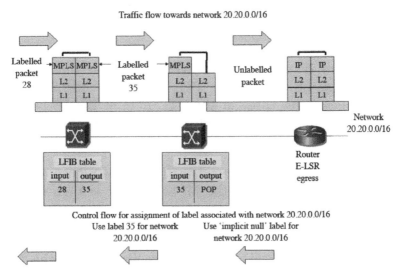

Figure 7.4. *PHP function*

The label search operation in the LFIB is then unnecessary. The *egress Edge LSR* router announces the *implicit-null* label to its router's upstream. An LSR with the *implicit-null* output label removes the packet label.

7.1.4. *MPLS header format*

A MPLS header occupies 4 bytes and contains the following fields (Figure 7.5):

– *Label*: this field includes the value of the label assigned to the packet;

– Exp (Experimental): this field includes the value of the traffic class to which the packet belongs;

– S (Stack): this field serves to stack MPLS headers. When this bit is set at one, the bottom of the pile has been reached and the level 3 header is placed just after it;

– TTL (Time to Live): this field has the same role as the TTL field in the IP header.

When an IP packet is labeled for the first time, in the *ingress Edge* LSR router the TTL field of the MPLS header can be set at the value of the TTL field of the IP header, which must first be decreased by 1 unit in accordance with the rule.

When a label is exchanged in the *Core* LSR router, the value of the TTL field of the MPLS header is then in turn decreased by 1 unit.

When a label is removed in the *egress Edge* LSR header, and the resulting label stack is empty, the value of the TTL field of the IP header must be replaced with the value of the TTL field of the MPLS header.

However, situations can occur where it is preferable to decrease the value of the TTL field of the IP header by 1 unit when the packet crosses an MPLS domain.

```
 0                   1                   2                   3
 0 1 2 3 4 5 6 7 8 9 0 1 2 3 4 5 6 7 8 9 0 1 2 3 4 5 6 7 8 9 0 1
+-+-+-+-+-+-+-+-+-+-+-+-+-+-+-+-+-+-+-+-+-+-+-+-+-+-+-+-+-+-+-+-+
|                   LABEL               | Exp |S|      TTL      |
+-+-+-+-+-+-+-+-+-+-+-+-+-+-+-+-+-+-+-+-+-+-+-+-+-+-+-+-+-+-+-+-+
```

Figure 7.5. *MPLS header format*

7.1.5. *DiffServ support*

In an MPLS network supporting the DiffServ mechanism, LSR routers execute traffic processing mechanisms depending on the value of the EXP field. The MPLS network introduces two types of LSP virtual circuits:

– the Label-inferred-class-LSP (L-LSP) virtual circuit transports only one traffic class;

– the EXP-inferred-class-LSP (E-LSP) virtual circuit transports several traffic classes simultaneously.

For the L-LSP model, LSR routers use the label and EXP field of the labeled packet to determine the per hop behavior (PHB) of the LSR router. Therefore, it is possible to have biunivocal correspondence between the DiffServ Code Point (DSCP) field value of the IP header and the value of the label representative of the traffic class and EXP field representative of the level of precedence. Thirteen labels are required to code all of the values of the DSCP field (Table 7.1).

DSCP field	Traffic group	Label Traffic group	EXP field Precedence
EF	EF	L1	000
AF43			011
AF42	AF4	L2	010
AF41			001
AF33			011
AF32	AF3	L3	010
AF31			001
AF23			011
AF22	AF2	L4	010
AF21			001
AF13			011
AF12	AF1	L5	010
AF11			001
CS7	CS7	L6	000
CS6	CS6	L7	000
CS5	CS5	L8	000
CS4	CS4	L9	000
CS3	CS3	L10	000
CS2	CS2	L11	000
CS1	CS1	L12	000
CS0	CS0	L13	000

Table 7.1. *Correspondence for L-LSP and EXP*

For the E-LSP model, LSR routers use the EXP field of the MPLS header to determine the PHB behavior of the LSR router. The EXP field contains 3 bits. Therefore, an E-LSP virtual circuit can code only eight traffic classes. The value of the EXP field is defined at the ingress of the MPLS network via a correspondence with the DSCP field of the IP header.

7.2. LDP protocol

7.2.1. *Principles of functioning*

The LDP enables the exchange of messages between adjacent LSR routers for the assignment of a label to an FEC. There are four types of LDP messages:

– discovery message, which detects adjacent LSR routers;

– initialization message, which establishes the LDP session between two LSR routers;

– announcement message, which provides correspondence between the label and the FEC;

– notification message, which signals errors.

First, the LSR router generates discovery messages. These messages are encapsulated by a User Datagram Protocol (UDP) header. The UDP segment is encapsulated by an IP header, the destination address of which is the multicast address 224.0.0.2. The adjacent LSR router also responds with a discovery message.

During exchanges of discovery messages, each LSR router learns the identifier of adjacent routers. The LSR router with the strongest identifier is the active LSR router of each pair that opens the Transmission Control Protocol (TCP) connection, with the destination port number being 646.

The active LSR router then sends an initialization message containing the parameters of the LDP protocol. It waits for 15 s for the adjacent router to respond. If the response is negative, it resends the message with new parameters and waits 30 s for the adjacent router to respond. At each negative response, the active router doubles the wait time until it reaches 2 min.

The downstream label distribution method indicates that the propagation of LDP messages occurs from the nearest router from the target network to the most distant router (downstream to upstream). The downstream method has two variants: *Unsolicited Downstream* and *Downstream On Demand* (Figure 7.6). In the first variant, LSR routers located downstream systematically propagate all of their labels to their upstream neighbors. In the second variant, LSR routers located upstream explicitly request that LSR routers located downstream provide them with a label for the IP network requested.

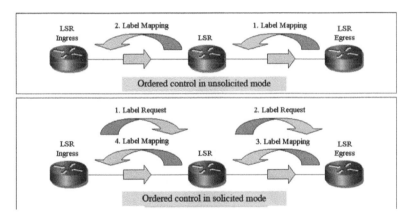

Figure 7.6. *Establishment of LSP virtual circuit*

The establishment of LSP virtual circuits can be controlled in an independent or ordered manner. In independent control mode, each LSR router freely decides to link a label to an FEC group and distributes this correspondence to its neighbors. In ordered control mode, an

LSR only links a label to a specific FEC group if it is the *egress Edge* LSR or if it has already received correspondence between the label and the FEC from the next hop LSR.

If an LSR router supports liberal retention mode, it maintains all the correspondence between the label and the FEC that it receives from the adjacent LSR routers, whether these routers are located downstream or not (Figure 7.7).

If an LSR router supports conservation mode, it preserves only FEC-Label associations originating from routers located downstream, in accordance with the indication given by the routing table (Figure 7.7).

With liberal retention mode, the LSR router can transfer data as soon as the routing protocol has converged following a modification of the MPLS network. However, it must maintain a table with all announcements received.

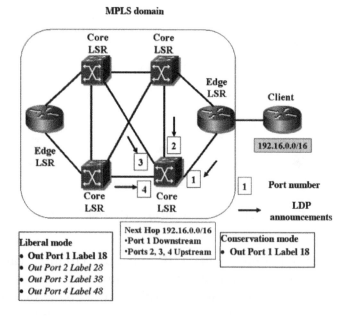

Figure 7.7. *Retention modes*

The LDP protocol implements two mechanisms used to detect loops. The first mechanism uses a counter increased by 1 unit when the LDP message crosses a router. When the maximum value is attained, the message is deleted. The second mechanism consists of registering the addresses of the routers crossed. If a router recognizes itself in the list, the message is deleted.

7.2.2. *LDP PDU format*

A protocol data unit (PDU) is composed of a header and one or more LDP messages. The header contains the following fields (Figure 7.8):

– *Version*: this field inserts the LDP protocol number and has a value of 1;

– *PDU Length*: this field inserts the length of the protocol data unit;

– *LDP Identifier*: this field inserts the router identifier (IP address) and the label space used by the router.

```
0                   1                   2                   3
0 1 2 3 4 5 6 7 8 9 0 1 2 3 4 5 6 7 8 9 0 1 2 3 4 5 6 7 8 9 0 1
+-+-+-+-+-+-+-+-+-+-+-+-+-+-+-+-+-+-+-+-+-+-+-+-+-+-+-+-+-+-+-+-+
|          Version          |           PDU Length            |
+-+-+-+-+-+-+-+-+-+-+-+-+-+-+-+-+-+-+-+-+-+-+-+-+-+-+-+-+-+-+-+-+
|                    LDP Identifier (router ID)                |
+-+-+-+-+-+-+-+-+-+-+-+-+-+-+-+-+-+-+-+-+-+-+-+-+-+-+-+-+-+-+-+-+
| LDP Identifier (label space) |
+-+-+-+-+-+-+-+-+-+-+-+-+-+-+-+-+
```

Figure 7.8. *LDP PDU header format*

Each type of LDP message contains the following fields (Figure 7.9):

– U (Unknown): upon reception of an unknown message:

    - if the bit is positioned at zero, the destination must send back a notification message,

- if the bit is positioned at one, the message is ignored;

– *Message Type*: this field inserts the message type (Table 7.2);

– *Message Length*: this field inserts the message size;

– *Message ID*: this field inserts the message identifier used by the receiver when it transmits the notification message pertaining to a received message;

– *Mandatory Parameters*: this block inserts the mandatory message parameters in TLV (Type, Length, Value) form;

– *Optional Parameters*: this block inserts optional message parameters in TLV form.

Message type	Description
HELLO	Discovery of adjacent LSR routers
INITIALIZATION	Establishment of LDP session (definition of parameters)
KEEPALIVE	Maintaining of session in the absence of message exchange
ADDRESS	IP address of next hop
ADDRESS_WITHDRAW	Removal of IP address of next hop
LABEL_MAPPING	Mapping between label and FEC
LABEL_REQUEST	Label request (*downstream on demand mode*)
LABEL_WITHDRAW	Removal of label by downstream router
LABEL_RELEASE	Removal of label by upstream router
LABEL_ABORT_REQUEST	Removal of label request in progress
NOTIFICATION	Notification of reception of a wrong message

Table 7.2. *Types of LDP messages*

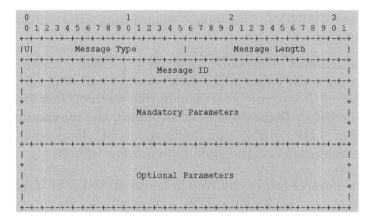

Figure 7.9. *Generic LDP message format*

7.2.3. *LDP messages*

LSR routers announce themselves to adjacent routers by sending a HELLO message that contains the following parameters:

– mandatory *hello* parameter: this parameter indicates the value of the *Hello Time* timer. This timer specifies the maximum waiting time for a HELLO message from the adjacent router. The default value is 15 s in the case of a multicast transmission and 45 s in the case of a unicast transmission. If the two routers announce two different values, the lower value is retained;

– optional *IPv4 address* parameter: this parameter contains the IP address of the LSR router, which determines the active router that initializes the TCP connection;

– optional *sequence number* parameter: increasing the sequence number initializes a new discovery if the previous discovery fails.

The INITIALIZATION message establishes an LDP session between two adjacent LSR routers. When the TCP session is established, the active router transmits an

INITIALIZATION message containing the following parameter:

– mandatory *session* parameter: this parameter determines the value of the LDP session maintaining timer (*Keepalive Time*), the label distribution method (*Downstream Unsolicited* or *Downstream On Demand*), the maximum size of the LDP PDU, an indication of loop detection implementation and the maximum number of hops.

The passive router can respond with an INITIALIZATION message or a KEEPALIVE message. If the required parameter is not accepted by the passive router, it responds with the NOTIFICATION error message.

Routers use the KEEPALIVE message to maintain the LDP session. This message contains no parameters. If no LDP message has been received when the *Keepalive Time* timer expires, the LDP session is released.

The ADDRESS message gives to the LSR router the router IP addresses of the next hop. It is transmitted after initialization of the LDP session or when a new address is configured. It contains the following parameters:

– mandatory *address list* parameter: this parameter determines the list of interface IP addresses of the downstream router.

The ADDRESS_WITHDRAW message enables the LSR router to remove the router IP addresses of the next hop. It contains the same parameter as the ADDRESS message.

The LABEL_MAPPING message is used by the router located downstream to give the upstream router the value of the label associated with a FEC. This message is sent by the downstream router at its own initiative (*Downstream Unsolicited* mode) or following a request from the upstream router (*Downstream On Demand* mode). The

LABEL_MAPPING   message   includes   the   following parameters:

– mandatory *FEC* parameter: this parameter contains the IP address of the network to reach for which a correspondence with the label is defined;

– mandatory *label* parameter: this parameter contains the value of the label associated with the FEC;

– optional *hop count* parameter: this parameter contains the number of hops executed by the message. The value is increased by 1 unit for each LSR router crossed. The limit value of the number of hops is defined by the *session* parameter of the INITIALIZATION message;

– optional *path vector* parameter: this parameter contains the list of IP addresses of the routers crossed. It is used to detect LDP message transmission loops;

– *request message* parameter: this parameter contains the identifier of the request expressed by the LABEL_REQUEST message.

The LABEL_REQUEST message is used in *Downstream On Demand* mode by the upstream router to request the value of a label associated with an FEC. It contains the mandatory *FEC* parameter and the optional *hop count* and *path vector* parameters.

The LABEL_WITHDRAW message is used by the downstream router to remove a label. It contains the mandatory *FEC* parameter and the optional *label* parameter.

The LABEL_RELEASE message is transmitted by the upstream router to cancel the label transmitted by the downstream router. It contains the mandatory *FEC* parameter and the optional *label* parameter.

The LABEL_ABORT_REQUEST message is used by the upstream router to cancel a LABEL_REQUEST request in progress. It contains the mandatory *FEC* and *request message* parameters.

Routers use the NOTIFICATION message to indicate a procedural error. If the error is fatal, the LDP session is released. The NOTIFICATION message contains the following parameters:

– mandatory *status* parameter: this parameter gives an indication of error processing (fatal error) and the type of error detected;

– optional *extended status* parameter: this parameter provides additional indications of the type of error;

– optional *returned PDU* parameter: this parameter provides a copy of the PDU that caused the error;

– optional *returned message* parameter: this parameter supplies a copy of the message that caused the error.

Table 7.3 summarizes the parameters associated with LDP messages.

7.3. VPN construction

7.3.1. *Network architecture*

The Virtual Private Network (VPN) function is introduced into the MPLS network via the following two functions, localized in the *Edge* LSR:

– partition of the *Edge* LSR routing table (RIB). One routing instance, called VPN Routing and Forwarding (VRF), is assigned to each interface of the *Edge* LSR, user side;

– marking of flows belonging to the same closed user group. This marking, corresponding to an MPLS label, is effective between two *Edge* LSRs.

Parameters	Messages	Required/optional
session	INITIALIZATION	mandatory
address list	ADDRESS	mandatory
	ADDRESS_WITHDRAW	mandatory
FEC	LABEL_MAPPING	mandatory
	LABEL_REQUEST	mandatory
	LABEL_WITHDRAW	mandatory
	LABEL_RELEASE	mandatory
	LABEL_ABORT_REQUEST	mandatory
label	LABEL_MAPPING	mandatory
	LABEL_WITHDRAW	optional
	LABEL_RELEASE	optional
hop count	LABEL_MAPPING	optional
	LABEL_REQUEST	optional
path vector	LABEL_MAPPING	optional
	LABEL_REQUEST	optional
request	LABEL_MAPPING	optional
message	LABEL_ABORT_REQUEST	mandatory
status	NOTIFICATION	mandatory
extended status	NOTIFICATION	optional
returned PDU	NOTIFICATION	optional
returned message	NOTIFICATION	optional

Table 7.3. *LDP message parameters*

The MPLS-VPN has the following components (Figure 7.10):

– customer edge (CE): this is generally the user router of the MPLS-VPN (aggregation network, another core network), which establishes adjacency with the MPLS-VPN edge router (provider edge (PE) router). It does not participate in packet labeling;

– PE: this is the new denomination of the *Edge* LSR router. It implements the VPN plus *Edge* LSR functions;

– P (Provider): this is the new denomination of the *Core* LSR router. It has no knowledge of the VPN. It functions identically to the *Core* LSR.

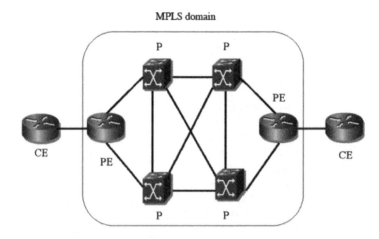

Figure 7.10. *MPLS-VPN network architecture*

The CE router transmits announcements concerning local routes to the PE router, and receives announcements concerning remote routes from the PE. Announcements are made using standard routing protocols.

The PE maintains VRF routing instances (Figure 7.11). A CE-side interface is associated with a VRF. Multiple CE-side interfaces can be grouped on one VRF. The PE informs the other PEs of the VPN routes constructed from the routes learned from the connected CE. A VPN route is an aggregate of the route and a Route Distinguisher (RD) identifier. Announcements between PEs are made using the Multiprotocol Border Gateway Protocol (MP-BGP).

The P router executes the label switch for the transfer of flows between two PEs. Label switching is executed using an LSP label that is different from the VPN label affected by

marking flows belonging to the same closed user group. The
P router processes no information concerning the VPN.

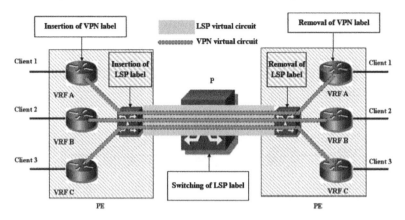

Figure 7.11. *VPN construction*

Label switching uses the same principles described for the
MPLS network:

– Open Shortest Path First (OSPF) or Intermediate
System-to-Intermediate System (IS-IS) intradomain routing
protocols enable knowledge of routes to reach PEs. The FEC
is composed of the PE address;

– the LDP protocol is used to assign a label to this FEC.

This control plan data is exchanged between adjacent
routers (PE or P) of the MPLS-VPN and used to establish the
LSP virtual circuit.

The distribution of VPN routes and the assignment of
VPN label use the MP-BGP. This control plan data is
exchanged between PEs.

The PE router processes routing of the incoming IP packet
and two labelings (Figure 7.12):

– the first label carries the VPN mark;

– the second label contains the identifier that enables label switching.

The P router executes label switching, acting only on the second label.

The outgoing PE router removes the two labels and routes the IP packet.

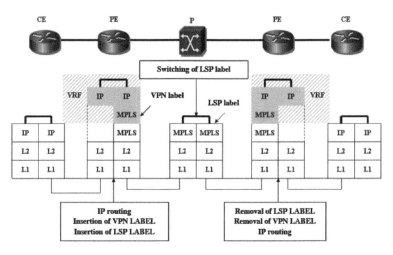

Figure 7.12. *MPLS-VPN network protocol architecture*

7.3.2. *Differentiation of routes*

The PE router must transfer the routes learned from the client to the other PEs via MP-BGP. It is offered the possibility of IP addresses overlapping between clients belonging to different VPNs. The client is even allowed to use private addresses. Typically, each VPN can use the 10.X.X.X. range of addresses.

The PE router has two types of RIB routing tables: the routing table specific to each VRF and the common routing table. Routing protocols, particularly MP-BGP, feed the common routing table in order to limit the number of instances. IP routes are differentiated by adding an RD identifier. Each VRF of a single VPN may or may not use the same RD. The RD identifier is not used to build a VPN, but only to differentiate between routes.

The RD identifier is a 64-bit field. A VPNv4 route is a 96 bit aggregation of an RD and an IPv4 route. The RD identifier is configured by VRF and constructed with the following two fields:

– a first 32 bit field that can be the autonomous system (AS) number or an IP address (the PE's, for example);

– a second 32 bit field inserting a value specific to the VRF.

7.3.3. *Route target*

VPN construction is based on the Route Target (RT). The RT structure is identical to the RD structure. The RT identifier defines the manner in which VPNv4 routes will be inserted into the VRFs. Each VRF is configured with two parameters:

– RT export: this parameter is transmitted in the MP-BGP message in association with the announced route. It constitutes an extended BGP community;

– RT import: this parameter acts as a filter for the insertion of a route learned via the MP-BGP protocol. The VRF confirms whether the RT export value received is equal to the RT import value.

Route targets enable the constitution of different VPN topologies. A mesh network is created by associating all of the sites two by two. To exchange routes, each PE establishes the same route import and export rule: RT import RT export for every PE (Figure 7.13).

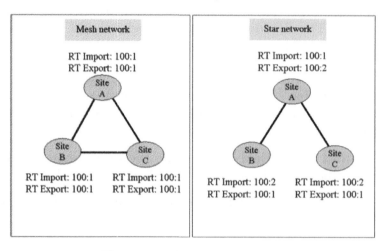

Figure 7.13. *VPN topology – 1/2*

A star network includes a central site (hub site) and peripheral sites (spoke sites). Route exchange takes place only between the central site and peripheral sites. PEs of peripheral sites do not have the same route import and export rules. The central-site PE establishes the following rules (Figure 7.13):

– RT import for the central site PE = RT export for the peripheral sites' PE;

– RT export for the PE central site = RT import for the peripheral sites' PE;

– RT export for the PE peripheral sites' PE ≠ RT import for the peripheral sites' PE.

More complex topologies can be obtained, for example the possibility of one site belonging to several VPNs (Figure 7.14).

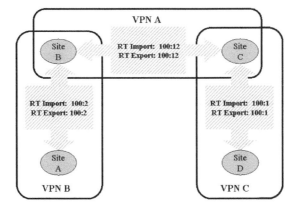

Figure 7.14. *VPN topology – 2/2*

7.3.4. *Principles of operation*

7.3.4.1. *Operation of control plane*

The MPLS-VPN implements the control plane for the construction of the LSP virtual circuit between the PE and the network isolation.

LSP construction is similar to the construction of a classic MPLS network. The difference lies in the identity of the FEC. In a classic MPLS network, the FEC is the address of the destination network. In an MPLS-VPN, the FEC is the address of the PEs. OSPF or IS-IS routing protocols enable the distribution of routes relative to the addresses of the PE routers.

In the example below, the address of the PE1 router is a private address (10.1.1.1). When the PE addresses are known for all of the PE and P routers, the LSP is constructed in the following manner (Figure 7.15):

– the PE1 router sends an LDP message to the P router assigning a value of 38 to FEC 10.1.1.1/32;

the P router sends an LDP message to the PE2 router assigning a value of 43 to FEC 10.1.1.1/32.

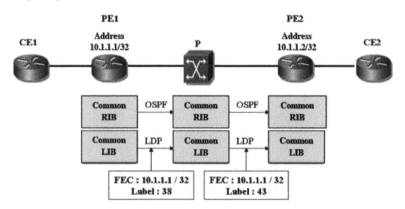

Figure 7.15. *Control plan operation – LSP construction*

VPN construction occurs in several stages (Figure 7.16):

– stage 1: the CE1 router announces route 10.1.1.0/24 to VRF A of router PE1. Identical addresses have been deliberately taken between clients attached to CE1 routers and routers in the MPLS-VPN, in order to show that there is no overlapping. The VRF updates its RIB table, transforms the IPv4 route into a VPNv4 route by aggregating the RD identifier 1:100 and assigns the label VPN 45;

– stage 2: VRF A of router PE1 transmits this information as well as the export RT identifier 1:101 to the common routing table, which updates the common LFIB;

– stage 3: the PE1 router transmits this information to the common routing table of the PE2 router in the MP-BGP message, which also contains the nest hop address 10.1.1.1 of the PE1 router;

– stage 4: based on the value of the 1:101 export RT target route, the common RIB transmits the MP-BGP announcement to VRF A of router PE2;

– stage 5: VRF A of router PE2 removes the RD identifier of the VPNv4 route, updates its RIB table and transfers the 10.1.1.0/24 network address to the CE2 router.

The common RIB table of the PE routers contains the 10.1.1.1/32 address of the PE1 router and the 1:100:10.1.1.0/24 address of the network connected to the CE1 router. The RD identifier 1:100 eliminates any ambiguity.

The RIB table dedicated to a VRF does not know the affected IP addresses of the PE and P routers in the MPLS-VPN.

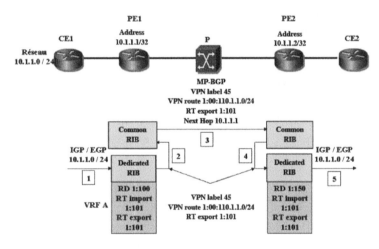

Figure 7.16. *Control plan functioning – VPN construction*

7.3.4.2. *Operation of traffic plane*

The transfer of traffic plane data between CE1 and CE2 routers initiates the following operations (Figure 7.17):

– the CE2 router transmits to the PE2 router an IP packet with a destination address of 10.1.1.1. This address is not the PE1 router's address, but the address of a user connected to the CE1 router;

– the VRF A of the PE2 router executes VPN labeling with a value of 45 and transmits the labeled packet to the common LFIB indicating the destination IP address, the PE1 router's address (10.1.1.1). The common LFIB labels the packet a second time, assigning it a value of 43 (LSP label);

– the P label switch executes a label switch and replaces the received label value of 43 with a value of 38;

– the PE1 router removes the 38 label and uses the VPN label value (45) to determine the destination VRF. VRF A removes the VPN label and transmits the IP packet to the CE2 router.

The IP packet containing the destination address 10.1.1.1 is not sent to a user embedded in the PE2 router because this packet is labeled 45. VRF A removes the 45 label. The RIB routing table dedicated to the VRF A indicates for this destination address the output toward CE1. User 10.1.1.1 embedded in the PE1 router is not known in the RIB routing table dedicated to VRF A, but only in the common RIB routing table.

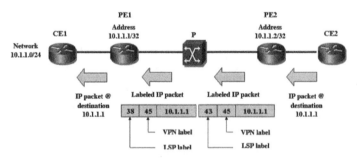

Figure 7.17. *Traffic plan functioning*

7.4. Network interconnection

Network interconnection consists of connecting two Internet Service Providers (ISPs) via the intermediary of an Internet transit provider.

Two scenarios can occur:

– the two ISPs belong to the same AS (identical AS numbers). Network interconnection is in hierarchical mode;

– the two ISPs do not belong to the same AS (different AS numbers). Interconnection is in recursive mode.

7.4.1. *Hierarchical mode*

The first stage must make it possible to transfer traffic across the different P routers of the Internet access networks (ISP network) and transit network. Several LSPs are established between the PEs using OSPF or IS-IS and LDP protocols (Figure 7.18):

– a first LSP virtual circuit is constructed between the access network PE1 router and the transit network PE2 router;

– a second LSP is constructed between the transit network PE2 and PE3 routers;

– a third LSP is constructed between the transit network PE3 router and the access network PE4 routers.

The second stage must make it possible to transfer the provider's traffic across the Internet transit network by isolating it. An ISP VPN label is created using the MP-interior BGP (MP-iBGP) protocol between PE2 and PE3 (Figure 7.18).

The third stage must make it possible to transfer client traffic across the Internet access and transit networks by isolating it. A client VPN label is created and the client route (10.2.0.0/16) is distributed using the MP-iBGP protocol between PE1 and PE4 (Figure 7.18).

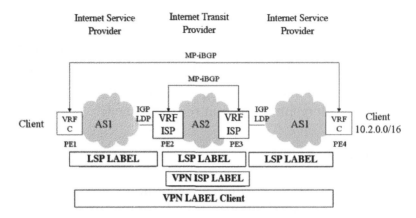

Figure 7.18. *Interconnection – hierarchical mode*

So that the PE1 and PE4 routers can communicate, the PE2 router must distribute the PE4 route toward PE1, and the PE3 router must distribute the PE1 route toward PE4.

7.4.2. *Recursive mode*

Because the Internet transit network connects networks of different ASs, the routing protocol used between the Internet access networks and the transit network is an exterior BGP (eBGP) protocol, rather than an intradomain protocol, such as OSPF or IS-IS (Figure 7.19).

For the same reason, the protocol used between routers PE1 and PE4 to transfer the client route is MP-eBGP rather than MP-iBGP.

The PEs of the Internet transit network must create VRF common to two different clients: Internet access networks AS1 and AS3.

The first stage, which consists of establishing a LSP virtual circuit between the different PEs, is different from the one described for hierarchical mode. The LSP virtual circuit is created only inside the Internet access network's AS and is not extended toward the Internet transit networks' PE.

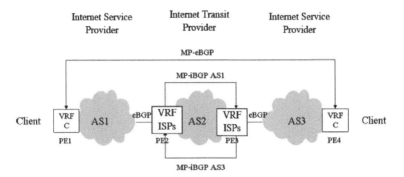

Figure 7.19. *Interconnection – recursive mode*

8

Ethernet VPN

8.1. Ethernet technology

Ethernet technology is employed in Local Area Networks (LANs) and Wide Area Network (WAN) aggregation networks. In LAN, two types of equipment can be deployed: hubs and switches. In aggregation networks, switches are the only equipment deployed.

Switches eliminate problems due to collision. They do not create a shared bus, but use buffer memories to store incoming frames. They also analyze destination Medium Access Network (MAC) addresses in order to transmit frames only on the relevant interface. The switching table contains information linking the MAC address and the egress interface. It is populated by a learning mechanism based on traffic analysis.

Each switch interface belongs to a separate collision domain. Each interface can function in full-duplex mode if a single user is located in the collision domain. An interface must function in half-duplex mode if multiple users are in the collision domain, for example when a hub is connected to a switch.

8.1.1. *Physical layer*

The Ethernet protocol distinguishes several physical layers, each with different characteristics in terms of rate or transmission support.

The 10Base-T physical layer uses two category-3 metallic pairs as a transmission support, with each pair supporting a direction of transmission. The principal transmission characteristics are as follows:

– the binary rate to be transmitted is equal to 10 Mbit/s;

– the maximum range is 100 m.

The 100Base-TX (fast Ethernet) physical layer uses two category-5 twisted pairs as a transmission support. The principal transmission characteristics are as follows:

– the binary rate to be transmitted is equal to 100 Mbit/s;

– the maximum range is 100 m.

Giga Ethernet technology uses the twisted pair (1000Base-T interface) and single-mode or multimode optical fibers as a transmission support for long distances (1000Base-LX interface) or short distances (1000Base-SX interface).

The 1000Base-T physical layer uses four category-5e or category-6 twisted pairs as a transmission support, with each pair enabling the two-way transmission of one quarter of the total rate. The two directions of transmission are separated using a differential transformer associated with an echo canceller. The principal transmission characteristics are as follows:

– the binary rate to be transmitted is equal to 1 Gbit/s. Each pair enables the two-way transmission of 250 Mbit/s;

– the maximum range is 100 m.

The 1000Base-SX physical layer uses two multimode optical fibers as a transmission support and functions at a wavelength of 850 nm, with each fiber assigned a direction of transmission. The principal transmission characteristics are as follows:

- the binary rate to be transmitted is equal to 1 Gbit/s;

- the maximum range is 550 m.

The 1000Base-LX physical layer uses two single-mode optical fibers as a transmission support and functions at a wavelength of 1,300 nm, with each fiber assigned a direction of transmission. The principal transmission characteristics are as follows:

- the binary rate to be transmitted is equal to 1 Gbit/s;

- the maximum range is 3 km.

The 10GBase physical layer functions only in full-duplex mode, on an optical fiber or metallic pair support.

The 10GBase-T physical layer uses four category-6a, category-6e, or category-7 twisted pairs as a transmission support.

Two letters are used to determine the type of interfaces on optical fibers (Table 8.1):

- the first letter determines the wavelength used;

- the second letter determines the type of line code.

1st letter	Type of wavelength	2nd letter	Type of line code
S	850 nm	X	8B/10B
L	1,310 nm	R	64B/66B
E	1,550 nm	W	SDH (STM-64)

Table 8.1. *Nomenclature of Ethernet interface at 10 Gbit / s*

10GBase-SR and 10GBase-SW interfaces use multimode optical fibers and a wavelength of 850 nm. They provide a range of up to 300 m.

10GBase-LR and 10GBase-LW interfaces use single-mode optical fibers and a wavelength of 1,310 nm. They provide a range of up to 10 km.

10GBase-ER and 10GBase-EW interfaces use single-mode optical fibers and a wavelength of 1,550 nm. They provide a range of up to 40 km.

The 10GBase-LX4 interface makes it possible to multiplex four wavelengths on one optical fiber. They use single-mode or multimode optical fibers and provide a range of up to 300 m for multimode fibers and 10 km for single-mode fibers.

8.1.2. *MAC layer*

There are several types of headers corresponding to the data link layer:

– the original Ethernet DIX (DEC, Intel and Xerox) header;

– the 802.3 MAC header completed by an 802.2 logical link control (LLC) header;

– the MAC/LLC header completed by a sub-network access protocol (SNAP) header.

The data link layer header contains the following fields (Figure 8.1):

Preamble: when the interface is functioning in half-duplex mode, this field contains 7 identical bytes (10101010) used for the time-synchronization of the receiver. In full-duplex functioning, this pattern is transmitted systematically in the absence of traffic.

Start of Frame Delimiter (SFD): this field contains the frame synchronization word (10101011).

MAC address: two fields are used to code the source and destination MAC addresses on 48 bits. The source MAC address field contains only one unicast MAC address. The destination MAC address field contains a unicast, multicast, or broadcast MAC address. MAC addresses are expressed in hexadecimal notation.

A unicast address is composed of the following fields:

– an individual/group (I/G) bit indicating whether the address is unicast (bit at zero) or multicast or broadcast (bit at one);

– a universal/local (U/L) bit indicating whether the address is administered locally (bit at one) or not (bit at zero);

– when the address is not administered locally, the next 22 bits correspond to a value assigned to the network card supplier by the IEEE organization;

– the remaining 24 bits represent the series number of the network card, assigned by the supplier.

A broadcast address has all bits at one, or FF-FF-FF-FF-FF-FF in hexadecimal notation.

A multicast address can be derived from a multicast IP version 4 (IPv4) address. The first three bytes have a fixed value of 01-00-5E. The 23 least significant bits of the multicast IP address are mapped on the last three bytes of the multicast MAC address, with the 24th bit set at zero.

Type (DIX): this field contains the identification of the data encapsulated by the Ethernet header. Its value is greater than 1,500.

Length (802.3): this field contains the size of the data encapsulated by the Ethernet header. Its value is less than or equal to 1,500.

Destination Service Access Point (DSAP) (802.2): this field contains the destination's service access point; it is not used and has a set value.

Source Service Access Point (SSAP) (802.2): this field contains the identification of the highest-level protocol that generated the data encapsulated by the LLC header. It corresponds to the *type* field of the DIX frame. This field is not used, given the limitation imposed by its size. The *type* field of the DIX frame is reported to the additional SNAP header.

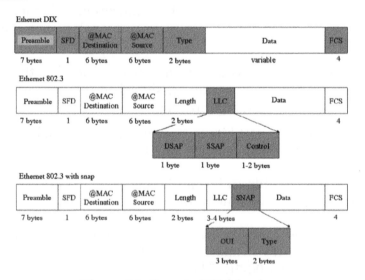

Figure 8.1. *Format of MAC headers*

Control (802.2): this field defines the functioning mode of the LLC protocol:

– LLC1: connectionless mode, without acknowledgment;

– LLC2: connected mode, with acknowledgment;

– LLC3: connectionless mode, with acknowledgment.

Organizationally Unique Identifier (OUI): this field is part of the SNAP header; it is usually not used and contains the code of the Ethernet card manufacturer.

Type (SNAP): this field contains the identification of the data encapsulated by the source.

Frame Check Sequence (FCS): this field contains a cyclic redundancy code used to detect errors on the Ethernet frame. Frames with errors detected are deleted.

Two consecutive frames are separated by a 96-bit Inter-Frame Gap (IFG) interval.

8.1.3. *VLAN isolation*

The Ethernet protocol has defined an additional four-byte field (Figure 8.2) used to execute a virtual LAN (VLAN) marking of the frame:

– two bytes insert the Tag Protocol Identification (TPID) field, used to indicate that VLAN marking has been added (value of 8,100 in hexadecimal notation);

– two bytes insert the Tag Control Identification (TCI) field.

The TCI field contains the following information:

– Priority Code Point (PCP) field, coded on three bits (Table 8.2), which provides the priority level of the frame. As for the DiffServ mechanism, these three bits are used to express service quality mechanisms associated with data flows in the switch;

– the canonical format indicator (CFI) field, coded on one bit. A switch always sets this value at 0;

– the VLAN Identification (VID) field, coded on 12 bits. It identifies the VLAN to which the frame source belongs. The VID field is also called Q-VLAN.

Priority level	Acronym	Traffic group
1	BK	Background
2	–	–
0	BE	Best effort
3	EE	Excellent effort
4	CL	Controlled load
5	VI	Video
6	VO	Voice
7	NC	Network control

Table 8.2. *Classification of traffic types*

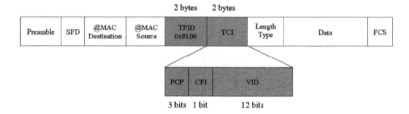

Figure 8.2. *Ethernet header marking*

Q-VLAN marking is used to isolate flows inside the LAN network to make up a Virtual Private Network (VPN). This isolation can be achieved in several ways, depending on the identification of the source:

– based on the port number;

– based on the source's MAC address;

– based on the source's IP address.

A VLAN table is composed of each switch (Figure 8.3). This table contains (for example), for each VLAN, the ports that are affected for them. The switch interfaces connecting users are access ports. It is usually forbidden to execute VLAN marking on this type of access.

Links between switches are called trunk ports. The switch executes VLAN marking on these interfaces. The mark indicates to which VLAN the frame source belongs.

When the S1 switch receives a frame on an access port, it recovers the destination MAC address and consults the switching table to determine the egress port.

If the egress port is an access port, the S1 switch consults the VLAN table to determine if the source and the destination belong to the same group. If the result is negative, the S1 switch deletes the frame. If the result is positive, it transfers the frame.

If the egress port is a trunk port, the S1 switch transfers the frame to the S2 switch, indicating the group to which the source belongs in the VLAN mark. When the S2 switch receives the frame, it recovers the destination MAC address and consults its switching table. If the egress port is an access port, the S2 switch consults the VLAN table to recover the group to which the destination belongs, and then compares it to the one contained in the VLAN mark of the frame received. If the result is negative, it deletes the frame. If the result is positive, it transfers the frame.

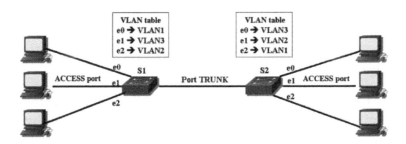

Figure 8.3. *LAN network isolation*

The VLAN table can be configured dynamically. During the user access control procedure by the 802.1x mechanism described in Chapter 2, the Remote Authentication Dial-In User Service (RADIUS) server authorizes the unblocking of the Ethernet port, if the user is authenticated, by sending a RADIUS *access-accept* message. This message may contain an attribute specifying the VLAN number assigned to this user.

8.2. PBT technology

The aggregation network of the WAN network is constructed from Provider Bridge (PB) switches that carry out the classic functions of an Ethernet switch (population of the switching table via learning; constitution of a spanning tree) and introduce a double VLAN mark.

At the aggregation network ingress, the Ethernet frame, Q-VLAN marked or not, receives a new service TAG (S-TAG) mark. The Q-VLAN marking changes denomination and becomes customer TAG (C-TAG). The S-TAG mark enables the operator to provide client flow isolation mechanisms and to preserve, in the aggregation network, the C-TAG marking executed by the client (Figure 8.4).

Likewise, the source (SA) and destination (DA) MAC addresses also receive a new denomination, customer SA (C-SA) and customer DA (C-DA).

The Tag Protocol Identification (TPID) field is used to indicate that S-TAG marking has been added (value of 88A8 in hexadecimal notation).

S-TAG marking is identical to C-TAG marking, with one exception: the CFI bit is replaced by the Drop Eligible Indicator (DEI) bit. This bit defines a level of precedence. If the bit has a value equal to one, it has a lower level of precedence (Figure 8.4).

The priority level coded in the PCP field addresses the scheduling mode for the queues. The precedence level coded in the DEI field is used for the congestion-avoidance mechanism.

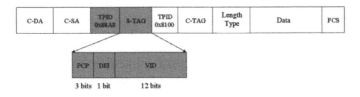

Figure 8.4. *Q-in-Q Ethernet frame structure*

Priority level and precedence level can also be coded in the PCP field only. The eight values of the PCP field can be used in different ways (Table 8.3):

– eight priority levels and zero precedence levels (8P0D);

– seven priority levels and a precedence level for one priority level (7P1D);

– six priority levels and a precedence level for two priority levels (6P2D);

– five priority levels and a precedence level for three priority levels (5P3D).

Priority	7		6		5		4		3		2		1		0	
Precedence	0	1	0	1	0	1	0	1	0	1	0	1	0	1	0	1
P C P	8P0D	7		6		5		4		3		2		1		0	
	7P1D	7		6		5	4	5	4	3		2		1		0	
	6P2D	7		6		5	4	5	4	3	2	3	2	1		0	
	5P3D	7		6		5	4	5	4	3	2	3	2	1	0	1	0

Table 8.3. *PCP field coding*

8.3. VPLS technology

8.3.1. *Network architecture*

VPLS technology is deployed in the core network. The reference architecture is modeled on that of the Multi-Protocol Label Switching (MPLS-VPN) network (Figure 8.5):

– the Provider Edge (PE) equipment hosts a Virtual Switch Instance (VSI) instead of the VPN Routing and Forwarding (VRF) virtual router;

– the Provider (P) equipment provides the same label switching function, no matter what type of PE.

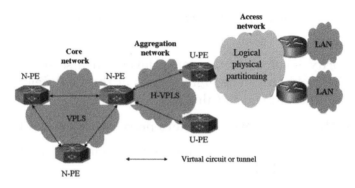

Figure 8.5. *VPLS network architecture*

H-VPLS architecture makes it possible to extend MPLS technology to the aggregation network. PE equipment is deployed at two levels (Figure 8.5):

– network-facing PE (N-PE), between the core network and the aggregation network;

– user-facing PE (U-PE), between the aggregation network and the access network.

The interconnection between N-PE equipments is a mesh of virtual circuits or tunnels, with the N-PEs connected pair to pair. The tunnel is the equivalent of the label switch path (LSP) virtual circuit of the MPLS network.

The interconnection between U-PE and N-PE equipment is a star of virtual circuits or tunnels, with the U-PEs connected to a single N-PE. This arrangement enables an optimum number of virtual circuits.

The PE equipment executes switching of the incoming Ethernet frame and a double labeling (Figure 8.6):

– the first Virtual Channel Label (VCL) labeling includes the VPN mark to create a Pseudo-Wire (PW) pseudo-link between two VSI virtual switches hosted in the PEs;

– the second Tunnel Label (TL) labeling contains the identifier allowing label switching in P equipment and the constitution of the tunnel.

The P equipment executes the label switch, acting only on the second label (TL) (Figure 8.6).

The egress PE equipment removes both labels, switches the Ethernet frame, and transfers it toward the aggregation network (Figure 8.6).

Label switching uses the principles described for the MPLS network:

– the Open Shortest Path First (OSPF) or Intermediate System to Intermediate System (IS-IS) protocols are used to learn the routes to the PEs. The FEC is composed of the PE's IP address;

– the Label Distribution Protocol (LDP) is used to assign the TL label to this FEC. The TL label is transported by part of the Ethernet over MPLS (EoMPLS) header, which is identical to a MPLS header and interpreted as such by the P equipment. The assignment of the TL label is executed via exchanges of LDP messages between adjacent P or PE equipment.

The VSI virtual switch has two types of interfaces:

– a PW interface on the VPLS network side;

– an Ethernet Attachment Circuit (AC) interface on the client side. The Ethernet header may contain no VLAN marks, or one or two marks.

Correspondence between the physical access point or AC interface VLAN mark number on the one hand, and the PW interface on the other hand, is configured manually.

The VCL label is associated with the PW pseudo-link. It is transported by the EoMPLS header and replaces, in the case of a Q-in-Q AC interface, the second VLAN mark, which is then eliminated.

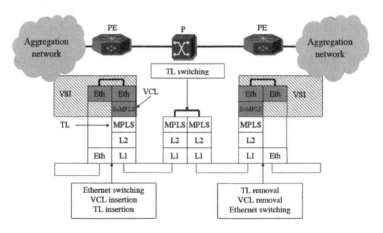

Figure 8.6. *Protocol architecture of VPLS network*

The PW pseudo-link between two VSI virtual switches is established either manually or dynamically using LDP messages exchanged between PEs (Figure 8.7).

The switching table of the VSI virtual switch is populated by learning, as for an Ethernet switch. Conversely, the functioning mode of the virtual switch is different with regard to the broadcasting of frames.

In a network of Ethernet switches, the Spanning Tree Protocol (STP) is used to create a spanning tree and to avoid network flooding by broadcast or multicast frames.

A different mechanism, split horizon, is used in the network composed of VSI virtual switches. When a VSI virtual switch receives a frame broadcast on a PW access point, it forbids its broadcasting on other PW access points and authorizes it on AC interfaces.

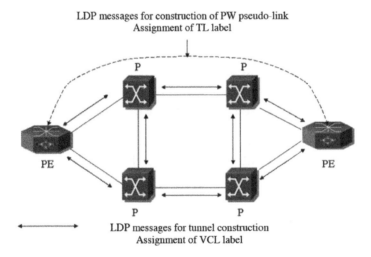

Figure 8.7. *Exchanges of LDP messages*

8.3.2. *EoMPLS header*

The MPLS header encapsulates an Ethernet frame whose header is stripped of the preamble, Start of Frame Delimiter (SFD) and Frame Checks Sequence (FCS) fields.

The EoMPLS header is composed of three words of 32 bits (Figure 8.8):

– the first two words each correspond to a MPLS header;

– the third word is a control word.

The first word contains the value dynamically assigned to the TL identifier by the LDP. The EXP field can be used to manage priorities. In this case, its value can be obtained by establishing correspondence with the Priority Code Point (PCP) field of the Ethernet header. The Stack (S) bit is positioned at zero to indicate that the bottom of the label stack has not been reached. The value of the Time To Live (TTL) field is set by the PE inserting the TL. This field is decreased by one unit for each P crossed.

The second word contains the value assigned to the VCL identifier. The EXP field is not used. The S bit is positioned at one to indicate that the bottom of the label stack has been reached. The value of the TTL field is set by the PE at two. This value is decreased only by the PE removing the VCL.

The third word contains a four-bit field at zero. The *sequence number* field, the use of which is optional, is used to restore Ethernet frames in sequence at output and to remove duplicate frames.

```
 0                   1                   2                   3
 0 1 2 3 4 5 6 7 8 9 0 1 2 3 4 5 6 7 8 9 0 1 2 3 4 5 6 7 8 9 0 1
+-+-+-+-+-+-+-+-+-+-+-+-+-+-+-+-+-+-+-+-+-+-+-+-+-+-+-+-+-+-+-+-+
|              Tunnel LABEL             | Exp |0|    TTL=255    |
+-+-+-+-+-+-+-+-+-+-+-+-+-+-+-+-+-+-+-+-+-+-+-+-+-+-+-+-+-+-+-+-+
|          Virtual Channel LABEL        | Exp |1|    TTL=2      |
+-+-+-+-+-+-+-+-+-+-+-+-+-+-+-+-+-+-+-+-+-+-+-+-+-+-+-+-+-+-+-+-+
|0 0 0 0|     Reserved          |       Sequence Number         |
+-+-+-+-+-+-+-+-+-+-+-+-+-+-+-+-+-+-+-+-+-+-+-+-+-+-+-+-+-+-+-+-+
```

Figure 8.8. *EoMPLS header format*

8.3.3. *LDP*

Changes made to the LDP pertain to Type Length Value (TLV) parameters, messages remaining unmodified. The pseudo-link establishment procedure is as follows:

1) A PW interface is configured for each PE, as well as its correspondence with the AC interface.

2) Each PE initiates a LDP session with a remote PE. The PEs exchange HELLO discovery messages and INITIALIZATION messages to establish the LDP session.

3) Each PE allocates a label (VCL) and an identifier to the PW pseudo-link. The VCL label is positioned in the *label* parameter; and the PW identifier in the *FEC* parameter. These two parameters are transmitted in a LABEL_MAPPING message.

8.3.3.1. *FEC parameter*

The structure of the *FEC* parameter contains the following fields (Figure 8.9):

PWid: this field identifies the type of FEC. It has a value of 80 in hexadecimal notation for this type of FEC described.

C-Bit: this bit indicates whether the control word in the EoMPLS header is used (value at one) or not (value at zero).

PW Type: this field indicates the type of PW pseudo-link. It has the following hexadecimal values:

– 0004 if the AC interface is a VLAN Ethernet interface;

– 0005 if the AC interface is an Ethernet interface.

PW info length: this field indicates the size of the pseudo-link identifier (PW ID) field and the TLV sub-parameter interface parameters.

Group ID: this field codes a value assigned to a set of PW pseudo-links.

PW ID: this field codes the value of the VCL label.

Interface parameters: this sub-parameter has a TLV structure providing information on the AC interface, such as the MTU size for example.

```
 0                   1                   2                   3
 0 1 2 3 4 5 6 7 8 9 0 1 2 3 4 5 6 7 8 9 0 1 2 3 4 5 6 7 8 9 0 1
+-+-+-+-+-+-+-+-+-+-+-+-+-+-+-+-+-+-+-+-+-+-+-+-+-+-+-+-+-+-+-+-+
|    PWid       |C|             PW Type         |PW info Length |
+-+-+-+-+-+-+-+-+-+-+-+-+-+-+-+-+-+-+-+-+-+-+-+-+-+-+-+-+-+-+-+-+
|                            Group ID                           |
+-+-+-+-+-+-+-+-+-+-+-+-+-+-+-+-+-+-+-+-+-+-+-+-+-+-+-+-+-+-+-+-+
|                             PW ID                             |
+-+-+-+-+-+-+-+-+-+-+-+-+-+-+-+-+-+-+-+-+-+-+-+-+-+-+-+-+-+-+-+-+
|                                                               |
//                   Interface Parameters TLV                  //
|                                                               |
+-+-+-+-+-+-+-+-+-+-+-+-+-+-+-+-+-+-+-+-+-+-+-+-+-+-+-+-+-+-+-+-+
```

Figure 8.9. *FEC parameter format*

For this type of FEC, the same *PW ID* is used at the two end points of the pseudo-link. Another type of FEC makes it possible to have a different identifier for each end point. This type of FEC does not contain the *group ID* and *interface parameters* fields. If these functions are necessary they are provided by additional parameters.

Label removal is done via the LABEL_WITHDRAW message. It is optionally possible to remove a set of labels for the group of PW pseudo-links identified by the *group ID* field.

8.3.3.2. *PW status parameter*

PE equipment can use the *PW status* parameter transported in the NOTIFICATION message to inform the remote PE of the PW pseudo-link's status. The negotiation procedure for using the *PW status* parameter is as follows:

1) The PE supporting this function includes the *PW status* parameter in the LABEL_MAPPING message.

2) If the remote PE does not support this function, it responds with the LABEL_RELEASE message, displaying the release reason as *label withdraw PW status method not supported*.

If the LABEL_MAPPING message does not contain the *PW status* parameter, the function is not used.

8.4. L2TPv3 technology

The layer 2 tunneling protocol version 3 (L2TPv3) is used to create a PW pseudo-link between two end points and to transfer Ethernet frames across an IP network.

Control messages are used for the establishment, the maintenance and the release of connections and sessions. A connection corresponds to the establishment of a relationship between two end points. When the connection is established, each flow-transfer constitutes a session. It is possible to establish multiple connections between two end points.

8.4.1. *Data message*

Ethernet frames are encapsulated by two headers, L2TP *session* and L2-*specific sublayer*, to constitute a data message. The data message can be directly encapsulated by an IP header (the *protocol* field has a value of 115) or by a User Datagram Protocol (UDP) header. The end point that initiates the L2TPv3 connection uses the value 1,701 for the destination port; it can also use the same value for the source port.

When the data message is encapsulated by an IP header, the L2TP *session* header is composed of two fields whose value is defined during the establishment of the session (Figure 8.10):

Session ID: this field identifies the session between two end points. It has a local meaning. The two-directional session is recognized using two identifiers; one local and one remote.

Cookie: this field is optional. It is four or eight bytes in size and is used to protect against the accidental or deliberate insertion of data messages into the tunnel.

```
 0                   1                   2                   3
 0 1 2 3 4 5 6 7 8 9 0 1 2 3 4 5 6 7 8 9 0 1 2 3 4 5 6 7 8 9 0 1
+-+-+-+-+-+-+-+-+-+-+-+-+-+-+-+-+-+-+-+-+-+-+-+-+-+-+-+-+-+-+-+-+
|                          Session ID                          |
+-+-+-+-+-+-+-+-+-+-+-+-+-+-+-+-+-+-+-+-+-+-+-+-+-+-+-+-+-+-+-+-+
|                    Cookie (0, 32, 64 bits)                   |
+-+-+-+-+-+-+-+-+-+-+-+-+-+-+-+-+-+-+-+-+-+-+-+-+-+-+-+-+-+-+-+-+
```

Figure 8.10. *L2TP Session header format – IP encapsulation*

When the data message is encapsulated by a UDP header, the following fields are added (Figure 8.11):

T: this bit is set at zero to indicate that it is a data message.

Ver: this field indicates the L2TP protocol version and has a value of three.

```
 0                   1                   2                   3
 0 1 2 3 4 5 6 7 8 9 0 1 2 3 4 5 6 7 8 9 0 1 2 3 4 5 6 7 8 9 0 1
+-+-+-+-+-+-+-+-+-+-+-+-+-+-+-+-+-+-+-+-+-+-+-+-+-+-+-+-+-+-+-+-+
|T|x|x|x|x|x|x|x|x|x|x|x| Vers |            Reserved            |
+-+-+-+-+-+-+-+-+-+-+-+-+-+-+-+-+-+-+-+-+-+-+-+-+-+-+-+-+-+-+-+-+
|                          Session ID                          |
+-+-+-+-+-+-+-+-+-+-+-+-+-+-+-+-+-+-+-+-+-+-+-+-+-+-+-+-+-+-+-+-+
|                    Cookie (0, 32, 64 bits)                   |
+-+-+-+-+-+-+-+-+-+-+-+-+-+-+-+-+-+-+-+-+-+-+-+-+-+-+-+-+-+-+-+-+
```

Figure 8.11. *L2TP Session header format – UDP encapsulation*

The L2-*specific sublayer* header is optional. Its use is indicated during the establishment of the session. It is composed of the following fields (Figure 8.12):

S: this bit indicates whether the *sequence number* field is valid (value at one). In the opposite case, the *sequence number* field must be ignored.

Sequence number: this field numbers each data message and is used to detect a loss, desequencing, or duplication.

```
 0                   1                   2                   3
 0 1 2 3 4 5 6 7 8 9 0 1 2 3 4 5 6 7 8 9 0 1 2 3 4 5 6 7 8 9 0 1
+-+-+-+-+-+-+-+-+-+-+-+-+-+-+-+-+-+-+-+-+-+-+-+-+-+-+-+-+-+-+-+-+
|x|S|x|x|x|x|x|x|                    Sequence Number            |
+-+-+-+-+-+-+-+-+-+-+-+-+-+-+-+-+-+-+-+-+-+-+-+-+-+-+-+-+-+-+-+-+
```

Figure 8.12. *L2-specific sublayer header format*

8.4.2. *Control messages*

Control messages are optional and support connection- and session-management procedures. The control message has three parts:

– L2TP *session* header, identical to the data message header. All bits of the *session ID* field are set at zero;

– control message header;

– type of control message formed of Attribute Value Pair (AVP) modules.

The control message header is created using the following fields (Figure 8.13):

T: this bit is set at one to indicate that it is a control message.

L: this bit is set at one to indicate that the *length* field is present.

S: this bit is set at one to indicate the presence of the *Ns* and *Nr* fields.

Ver: this field indicates the L2TP protocol version and has a value of three.

Length: this field indicates the size of the control message starting at bit *T*.

Control connection ID: this field contains the local connection identifier. It is communicated to the remote end point during the establishment of the connection in an AVP.

Ns: this field contains the sequence number of the control message generated.

Nr: this field contains the sequence number of the next control message expected.

The *Ns* and *Nr* fields support the windowing mechanism and are used to detect the loss of control messages and for the retransmission.

```
 0                   1                   2                   3
 0 1 2 3 4 5 6 7 8 9 0 1 2 3 4 5 6 7 8 9 0 1 2 3 4 5 6 7 8 9 0 1
+-+-+-+-+-+-+-+-+-+-+-+-+-+-+-+-+-+-+-+-+-+-+-+-+-+-+-+-+-+-+-+-+
|T|L|x|x|S|x|x|x|x|x|x|x| Vers  |            Reserved           |
+-+-+-+-+-+-+-+-+-+-+-+-+-+-+-+-+-+-+-+-+-+-+-+-+-+-+-+-+-+-+-+-+
|                           Length                              |
+-+-+-+-+-+-+-+-+-+-+-+-+-+-+-+-+-+-+-+-+-+-+-+-+-+-+-+-+-+-+-+-+
|                     Control Connection ID                     |
+-+-+-+-+-+-+-+-+-+-+-+-+-+-+-+-+-+-+-+-+-+-+-+-+-+-+-+-+-+-+-+-+
|              Ns               |              Nr               |
+-+-+-+-+-+-+-+-+-+-+-+-+-+-+-+-+-+-+-+-+-+-+-+-+-+-+-+-+-+-+-+-+
```

Figure 8.13. *Control message header format*

The AVP module has the following structure (Figure 8.14):

M: when this bit is set at one, the receiver must release the connection or the session if the AVP module is not recognized.

H: this bit indicates to the receiver whether sensitive information in the AVP module is encrypted.

Length: this field contains the size of the AVP module.

```
 0                   1                   2                   3
 0 1 2 3 4 5 6 7 8 9 0 1 2 3 4 5 6 7 8 9 0 1 2 3 4 5 6 7 8 9 0 1
+-+-+-+-+-+-+-+-+-+-+-+-+-+-+-+-+-+-+-+-+-+-+-+-+-+-+-+-+-+-+-+-+
|M|H| res   |       Length      |          Vendor ID            |
+-+-+-+-+-+-+-+-+-+-+-+-+-+-+-+-+-+-+-+-+-+-+-+-+-+-+-+-+-+-+-+-+
|       Attribute Type          |                               |
+-+-+-+-+-+-+-+-+-+-+-+-+-+-+-+-+                               +
|                                                               |
//                     Attribute Value                         //
|                                                               |
+-+-+-+-+-+-+-+-+-+-+-+-+-+-+-+-+-+-+-+-+-+-+-+-+-+-+-+-+-+-+-+-+
```

Figure 8.14. *AVP module format*

Vendor ID: if all bits are set at zero, the AVP module is defined by the norm. In the opposite case, the field includes the vendor identifier to indicate that the AVP module is specific to the vendor.

Attribute type: this field contains the meaning of the AVP module.

Attribute value: this field contains the content of the AVP module.

8.4.3. Procedures

8.4.3.1. *Connection management*

Connection establishment begins with a three-exchange handshake:

1) When the L2TP mechanism has been activated at the router, the Start-Control-Connection-Request (SCCRQ) message is generated to initiate dialogue with the remote router and notify it of the capacities supported by the local router. The message contains the AVP module indicating the value of the *control connection ID* identifier retained by the local router (Table 8.4).

2) The remote router responds with a Start-Control-Connection-Reply (SCCRP) message containing its own capacities. This message indicates that the dialogue can continue. It contains the AVP module indicating the value of the *control connection ID* identifier retained by the remote router (Table 8.4).

The authentication and integrity checking of control messages are optional functionalities. The seal is transmitted in a specific AVP module. It is calculated by a hash function using the following elements:

– control message header;

– AVP modules;

– a local and remote random number, transmitted in an AVP during establishment of the connection;

– a secret shared between the two end points.

Upon reception of a control message, the router calculates a seal locally and compares it to the seal received. If the two seals match, the message is authenticated. In the opposite case, it is deleted.

Type of AVP	Presence of AVP	Meaning
Message type	Mandatory	Designation of control message
Host name		Router name
Router ID		Router identification (for example its IP address)
Assigned control connection ID		Value assigned to *control connection ID* parameter
Pseudo-wire capabilities list		Type of data handled
Random vector	Optional	Value used for encryption of sensitive data in AVP
Control message authentication nonce		Value of random number used for calculation of seal
Message digest		Value of seal
Control connection tie breaker		Value used to eliminate doubt when each end point opens a connection
Vendor name		Vendor name
Receive window size		Maximum value of reception window
Preferred language		Language used for fields coded in ASCII

Table 8.4. *Composition of SCCRQ and SCCRP messages*

3) The local router responds to the SCCRP message received with a Start-Control-Connection-Connected (SCCCN) message. This message indicates that the connection has been established.

The SCCCN message contains the following AVPs: *message type, random vector* and *message digest*.

When the connection is established, the two end points exchange HELLO messages at regular intervals to maintain the connection. If the HELLO message is not received during the hold time of the connection, the end point considers the connection to have been released and sends the Stop-Control-Connection-Notification (StopCCN) message.

The HELLO message contains the following AVPs: *message type*, *random vector* and *message digest*.

The StopCCN message contains the following AVPs: *message type*, *result code*, *random vector*, *message digest*, and *assigned control connection ID*. The *result code* AVP, the presence of which is mandatory, indicates the reason for the connection release. The presence of the *assigned control connection ID* AVP is mandatory if the StopCCN message follows the SCCRQ and SCCRP messages and optional otherwise.

8.4.3.2. Session management

Session management is optional and it can be initiated when the router receives an incoming call. This router provides the L2TP Access Concentrator (LAC) function (Figure 8.15).

Session establishment also starts with a three-exchange handshake:

1) When the interface is activated, the router transmits, in the Incoming-Call-Request (ICRQ) message, the values of the *session ID* and *cookie* parameters, as well as information concerning the use (or non-use) of the L2-*specific sublayer* header (Table 8.5).

2) Upon reception of the ICRQ message, the remote router responds with the Incoming-Call-Reply (ICRP) message containing its parameter values.

The ICRP message contains the following AVPs: message type, local session ID, remote session ID, circuit status, random vector, message digest, assigned cookie, L2-specific sublayer, data sequencing, Tx connect speed, Rx connect speed and physical channel ID.

3) Upon reception of the ICRQ message, the local router responds with the ICCN (Incoming-Call-Connected) message indicating that the session has been established.

Type of AVP	Presence of AVP	Meaning
Message type		Designation of control message
Local session ID		Session identifier assigned by local router
Remote session ID		Session identifier assigned by remote router
Serial number	Mandatory	Unique session identifier used for administrative purposes
Pseudo-wire type		Indication of type of data encapsulated
Remote end ID		Identifier used to link session to circuit
Circuit status		Indication of circuit status
Random vector		Value used for encryption of sensitive data in AVP
Message digest		Seal value
Assigned cookie		Value assigned to *cookie* field of L2TP *session* header
Session tie breaker		Value used to eliminate doubt when each end point opens a session
L2-specific sublayer	Optional	Indication of presence (or non-presence) of L2-*specific sublayer* header
Data sequencing		Indication of delivery of data in sequence or not
Tx connect speed		Rate of data generated by end point initializing session
Rx connect speed		Rate of data emitted by the other end point
Physical channel ID		Physical interface identifier

Table 8.5. *Composition of ICRQ message*

The ICCN message contains the following AVPs: message type, local session ID, remote session ID, random vector, message digest, L2-specific sublayer, data sequencing, Tx connect speed, Rx connect speed and circuit status.

A router can notify the remote end point of a modification of the characteristics of the PW pseudo-link via a Set-Link-Info (SLI) message.

A router can release a PW pseudo-link and inform the remote end point of this via the Circuit-Disconnect-Notify (CDN) message.

If the router is initiating the establishment of the call, the procedure is identical but uses messages pertaining to outgoing calls. This router provides the L2TP Network Server (LNS) function (Figure 8.15).

In summary, the routers at the end points of the L2TP link can execute the following functions (Figure 8.15):

– LAC – LAC. The two end point routers executing the LAC function receive an incoming call at one end point and terminate the incoming call at the remote end point;

– LAC – LNS. The end point router executing the LAC function receives an incoming call. The other router executing the LNS function terminates the incoming call and can initialize an outgoing call;

– LNS – LNS. Both routers execute the LNS function. One of the two routers must initialize an outgoing call. The other router can also execute the same operation.

In the event that an Ethernet frame is transferred across an IP network, the router must indicate, during establishment of the connection, the types of sessions that can be established, described in the *pseudo-wire capabilities list* AVP:

– a session between two Ethernet ports;

– a session between two Ethernet VLANs.

An association created between an Ethernet port or Ethernet VLAN, on the one hand and a PW pseudo-link, on the other hand, initializes the establishment of a session. The outgoing call method is used, with the end point routers executing a LNS function (Figure 8.15).

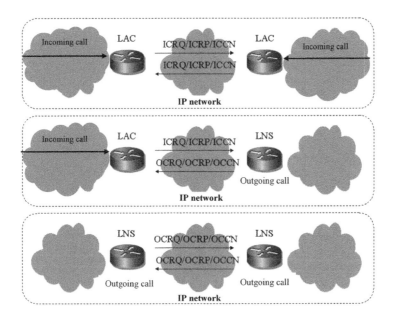

Figure 8.15. *Functions of L2TPv3 routers*

9

Firewalls

9.1. Technologies

Firewalls are entities positioned at the interconnection point between two networks, for example, at the interface between Local Area Networks (LANs) and Wide Area Networks (WANs).

The principal function of firewalls is to protect the LAN, considered as a trusted network in comparison to the WAN, which is a potential source of attacks. Only traffic explicitly authorized by the locally applied security policy is authorized to cross firewalls.

The functions provided by firewalls can be grouped into the following three areas:

– packet filter;

– Application-Level Gateway (ALG);

– Network Address Translation (NAT) and Network Address and Port Translation (NAPT).

0.1.1. *Packet filter*

Packet filtering is executed in two modes:

– stateless mode, in which packets are analyzed individually;

– stateful mode, in which the sequencing of packets in a single Transmission Control Protocol (TCP) connection is checked.

Stateless packet filtering is executed at the level of Internet Protocol (IP) and TCP or User Datagram Protocol (UDP) headers, using criteria applied to the following fields (Figure 9.1):

– the source and destination addresses (IP header);

– the *Protocol* field identifying the type of data encapsulated by the IP header;

– the source and destination port numbers (TCP or UDP header).

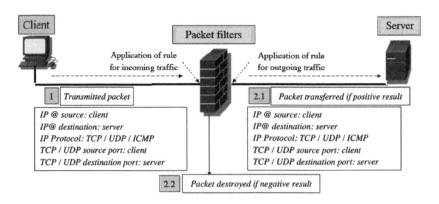

Figure 9.1. *Packet filtering*

Packet filtering is initialized using Access Control Lists (ACLs), according to the following three rules:

– everything that is authorized (*permit* command) is not forbidden; for example, the IP address 10.1.1.1 is authorized;

– everything that is forbidden (*deny* command) is not authorized; for example, the network 10.1.1.0/24 is forbidden. This prohibition does not include the 10.1.1.1 address, which has been previously authorized;

– everything that is not explicitly authorized is implicitly forbidden; for example, in this case, all other networks are forbidden.

Once the rule has been established, it must be applied to an interface indicating the direction of traffic; incoming or outgoing.

Stateful packet filtering completes stateless packet filtering due to the fact that it controls other fields of the TCP header (*sequence number* and *acknowledge number*) and verifies the compliance of exchanges with the TCP protocol state machine.

Figure 9.2 gives a simplified description of the states corresponding to the opening (SYN and ACK flags) and closing (FIN and ACK flags) of the TCP connection.

Stateful packet filtering must have a specific function for the UDP transport protocol. This protocol is stateless and does not implement a connection opening or closing process. For this type of packet, filtering can take place in the application. For example, a Domain Name System (DNS) response from the WAN is accepted only if packet filtering has previously detected the corresponding request.

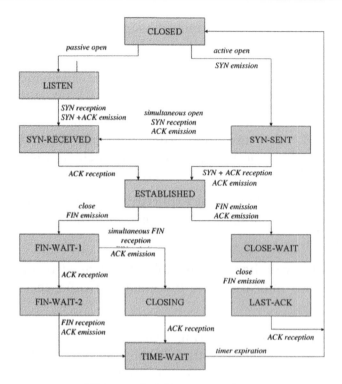

Figure 9.2. *TCP state machine*

9.1.2. *Applicative gateway*

The applicative gateway is complementary to packet filtering because it executes checks in the application layer. This function is usually provided by proxy or reverse proxy servers (Figure 9.3).

The TCP connection is established by the client by positioning the SYN bit. The final server accepts the opening by positioning the SYN bit in its response. The introduction of proxy or reverse proxy servers corresponds to the opening of two TCP connections; the first one is established between the client and proxy or reverse proxy server; the second one between the proxy or reverse server and final server.

The proxy server is used when the client is localized in the LAN and the final server is localized in the WAN. The client must be configured to transmit outgoing traffic systematically toward the proxy server.

Conversely, the client accesses the reverse proxy server when the final server is located in the LAN. The reverse proxy server must be configured to transmit incoming traffic toward the final server.

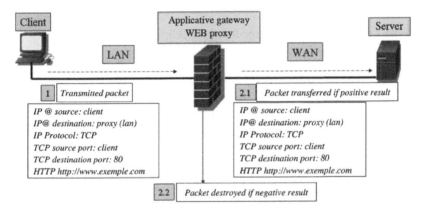

Figure 9.3. *Applicative gateway*

9.1.3. *NAT/NAPT device*

The IPv4 addressing plan defines the ranges of private addresses used in LAN:

– 10.X.X.X-type addresses for class A addresses;

– 172.16.X.X to 172.31.X.X-type addresses for class B addresses;

– 192.168.X.X-type addresses for class C addresses.

The NAT function is used to translate the private IP address of a LAN into a public IP address routable in a WAN. A public IP address is consumed for each private IP address (Figure 9.4).

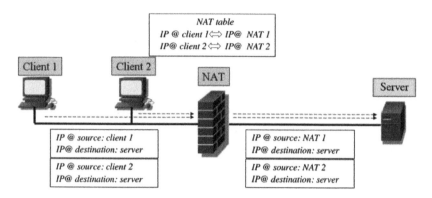

Figure 9.4. *NAT function*

The NAPT function enables the double translation of the private IP address and the port number in order to assign a single public IP address for multiple private addresses (Figure 9.5). The translation of the port number is crucial for the following two reasons:

– both sources can pick the same value for the port number;

– the source port number is the only parameter used to identify the session when the translation has been executed.

NAT or NAPT tables enable the opposite-direction relay of packets received from the server toward the client posts.

The NAT/NAPT mechanism is static when the NAT table is manually configured. This mechanism makes it possible to localize a server in the LAN, and therefore to receive a session request from the WAN.

The NAT/NAPT mechanism is dynamic when the table is populated after the establishment of the session by the client. This mechanism is used only for a client localized in the LAN and wishing to open a session with a public server.

The NAT/NAPT device protects hosts localized in the LAN. Any attempt to establish a session from the WAN can only succeed if the NAT/NAPT table has been previously populated.

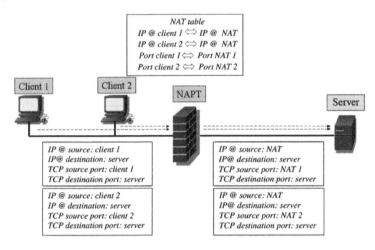

Figure 9.5. *NAPT function*

The NAT/NAPT device executes different types of filtering of packets transmitted by the WAN:

– full cone: the NAPT entity sends packets to the client, which has previously opened the session, without verifying the origin of the packet (Figure 9.6);

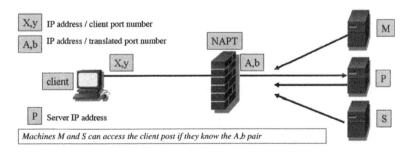

Figure 9.6. *Full cone*

– restricted cone: the NAPT entity sends packets to the client, which has previously opened the session, after verifying the origin of the packet. The restriction can affect the IP address alone (Figure 9.7) or the pair composed of the IP address and port number (port-restricted cone).

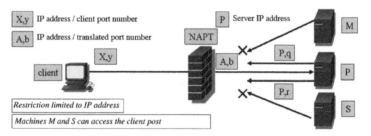

Figure 9.7. *Restricted cone*

– symmetric cone: this method consists of assigning an IP address/port number pair for each session established by the client. For packets received from public servers, the NAPT entity functions like a port-restricted cone (Figure 9.8).

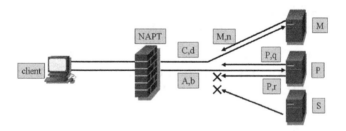

Figure 9.8. *Symmetric cone*

9.2. NAT/NAPT device crossing

The NAT/NAPT device presents difficulties for certain flows that must cross it:

– applications with a specific identification such as the Internet Control Message Protocol (ICMP);

– protocols protecting the IP packet load, such as, for example, the Encapsulated Security Payload (ESP) protocol of the IP Security (IPSec) mechanism;

– applications transporting IP addresses, such as, for example, Session Information Protocol (SIP) and Session Description Protocol (SDP);

– flows established dynamically, such as File Transfer Protocol (FTP) and Real-time Transport Protocol (RTP);

– fragmented IP packets.

9.2.1. *ICMP protocol*

The ICMP protocol has a specific characteristic of not using port numbers; ICMP messages are directly encapsulated by the IP header.

A specific method must be used for crossing the NAT device. Rather than relying on port numbers, the NAT device uses the *Identifier* field of the ICMP message. The entry in the NAT table must be associated with a timer, the value of which must be greater than 1 min.

As with the source port number translated by the NAPT device, the *Identifier* field is the only parameter that enables two sessions to be differentiated when the translation has been completed. The translation must include this field, since two sources can use the same value simultaneously for it.

Figure 9.9 shows the functions that the NAT device must execute in order to generate the ICMP *Echo Request* message and to receive the ICMP *Echo Reply* message.

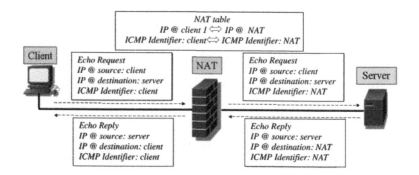

Figure 9.9. *Processing of ICMP Echo Request and Echo Reply messages*

Some ICMP error messages contain a copy of the IP packet that caused the error. The NAT/NAPT device must then execute the following operations:

– recover the IP address and source port number. These parameters are in the IP header of the packet that caused the error. This packet is included in the load of the ICMP error message;

– consult the NAPT table and translate these values so that the destination receives information consistent with the IP packet transmitted;

– translate the destination IP address of the IP header containing the ICMP error message.

Figure 9.10 shows the functions that the NAT/NAPT device must execute upon reception of an ICMP *Destination Unreachable* message. This message is returned by a router when it is unable to transfer a received IP packet.

9.2.2. *IPSec mechanism*

In transport mode or tunnel mode, the Authentication Header (AH) protocol authenticates the entire IP packet. Therefore, it is not possible in this case to modify the IP address of the source or destination, which makes the crossing of the NAT/NAPT device impossible.

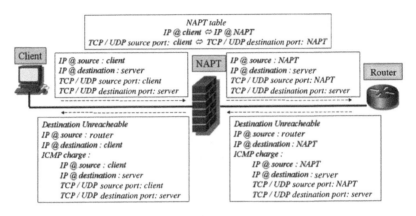

Figure 9.10. *Processing of ICMP Destination Unreachable message*

The ESP protocol of the IPSec mechanism protects the header of the TCP or UDP transport protocol but does not intervene with the packet's IP addresses. The crossing of the NAT/NAPT device by IP packets integrating this ESP header cannot be done successfully unless access to a transport protocol is possible to execute the following operations:

– modification of the IP address, with a consequent recalculation of the value of the *Checksum* field of the TCP or UDP header;

– translation of the source or destination port numbers by the NAPT device.

The NAT-T (Transversal) function adds a UDP header between the IP header and ESP header in order to make the previous two operations possible (Figure 9.11).

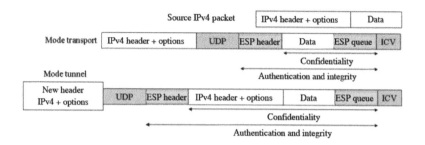

Figure 9.11. *ESP data encapsulation by UDP header*

The Internet Key Exchange (IKE) protocol must negotiate the NAT-T function and detect the presence of the NAT/NAPT device.

The negotiation of the NAT-T function is determined by the exchange of the VENDOR ID block containing the MD5 hash of the RFC 3947 character string, the value of which is equal to 4a131c81070358455c5728f20e95452f in hexadecimal notation.

The presence of the NAT/NAPT device is detected via the sending of the hash of the source and destination IP addresses and port numbers in two NAT-D (Discovery) blocks.

During the negotiation and detection phases, the UDP header encapsulating the IKE messages uses a value of 500 for the source and destination port numbers. When the presence of the NAT/NAPT device has been detected, the value of the source and destination port numbers is 4500.

The same port number (4500) is used for the UDP protocol encapsulating the IKE message or ESP protocol data unit (PDU). To be able to differentiate between the two types, a mark coded on four octets with all bits set at zero is inserted between the UDP header and IKE message.

In the case of transport mode, the destination must be able to verify the *Checksum* field of the UDP or TCP header, which has been calculated with the source IP address. Since this address has been translated by the NAT/NAPT device, the destination is informed by an IKE exchange indicating the source IP address in an NAT-OA (Original Address) block.

9.2.3. *SIP, SDP and RTP protocols*

The SIP protocol is used to execute the following operations:

– registration of the telephone post with the REGISTRAR entity;

– establishment of telephone communication.

SIP messages generated by a telephone post contain IP addresses corresponding to the private addressing plan of the LAN. For example, the telephone terminal generates an SIP REGISTER request when it connects to the LAN, in order to indicate its IP address to the REGISTRAR entity located in the WAN public network.

REGISTER sip:registrar.atlanta.com SIP/2.0

...

Contact: <sip:alice@**192.168.1.100**>

The SDP protocol is used in complement to the SIP protocol during the establishment of telephone communication:

– the SDP message is associated with the SIP INVITE request sent to the PROXY SERVER entity. This request is initiated by the post making the call.

– the SDP message is also associated with the temporary response SIP 1xx or with the definitive response 200 OK from the post being called.

SDP messages incorporate characteristics of RTP flow (IP addresses and port numbers of each end point), which must be established between the two telephone posts.

```
INVITE sip:bob@biloxi.com SIP/2.0
...

v=0
o=alice 2890844526 2890844526 IN IP4 192.168.1.100
s=-
c=IN IP4 192.168.1.100
t=0 0
m=audio 3456 RTP/AVP 97 96
a=rtpmap:97 AMR
a=rtpmap:96 telephone-event
```

Several mechanisms have been defined in order to enable the crossing of the NAT/NAPT device:

– session traversal utilities for NAT (STUN);

– traversal using relay NAT (TURN);

– interactive connectivity establishment (ICE).

9.2.3.1. *STUN protocol*

STUN messages are exchanged between a client (the telephone post) located in the private LAN and a STUN server located in the public WAN. The STUN server can be integrated with the REGISTRAR and PROXY SERVER entities.

The STUN protocol enables the client to detect the presence of an NAT/NAPT device, to determine the cone type (full cone, restricted cone and port-restricted cone), and to recover the IP address and port number used on the public network side. This operation must be executed separately for SIP and RTP flows (Figure 9.12).

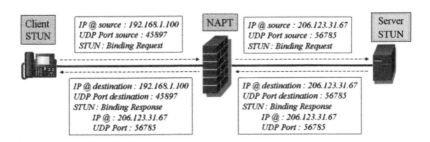

Figure 9.12. *STUN procedure*

The STUN protocol does not function with a symmetric NAT/NAPT device. Since the address of the STUN server is different from the address of the remote end point (as in the case of RTP flow), the mapping executed by the NAT/NAPT device will be different for the STUN server and the remote end point.

The STUN protocol does not constitute a solution in itself for crossing the NAT/NAPT device; rather it represents a tool to be used jointly with other mechanisms such as TURN or ICE.

9.2.3.2. *TURN protocol*

The TURN mechanism contributes a response for crossing the symmetric NAT/NAPT device by inserting a TURN server integrating an RTP flow relay entity. This provision makes it possible to obtain the same IP address for the TURN server and the remote end point of the RTP flow corresponding to the relay entity.

The TURN client (post A) transmits the ALLOCATE request to the TURN server, which responds with an ALLOCATE SUCCESS RESPONSE message containing the following information (Figure 9.13):

– IP address and port number of the relay entity;

– IP address and port number translated by the NAT/NAPT device.

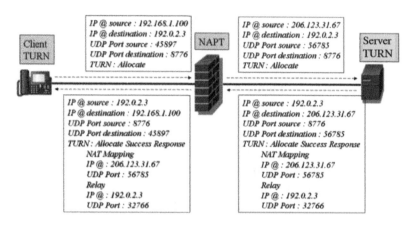

Figure 9.13. *TURN procedure – allocation phase*

When post A (the caller) generates its SIP INVITE request to post B (the callee), it indicates in the SDP message the IP address and port number allocated by the TURN server. In turn, post B supplies the parameters to receive the RTP flow (Figure 9.14).

Post A transmits the RTP flow to the relay entity of the TURN server, which forward the RTP flow to post B. The UDP segment containing the RTP flow is encapsulated by a TURN SEND REQUEST header in which post A supplies the IP address and port number of post B.

Post B transmits the RTP flow to the relay entity of the TURN server, which forward the RTP flow to post A. The UDP segment containing this RTP flow is encapsulated by a TURN DATA INDICATION header in which the TURN server supplies the IP address and port number of post B.

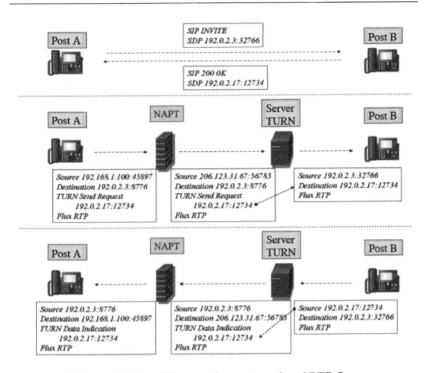

Figure 9.14. *TURN procedure – transfer of RTP flow*

9.2.3.3. *ICE mechanism*

The ICE mechanism enables two posts to agree on the IP address and port number used for the RTP flow. It is based on the use of the STUN and TURN protocols and introduces extensions of the SDP protocol.

During the inventory phase, post A (the caller) recovers the IP address and port number of the candidate interfaces (Figure 9.15). Three types of interfaces are defined:

– post interface (HOST);

– NAT/NAPT device interface (SERVER REFLEXIVE);

– TURN server relay entity interface (RELAYED).

Post A transmits to post B (the callee) the attributes of each candidate in an SDP message associated with an INVITE request. Post B executes the same operations as post A and transmits its candidates to post A in the SDP message associated with a temporary 1xx-type response (Figure 9.15).

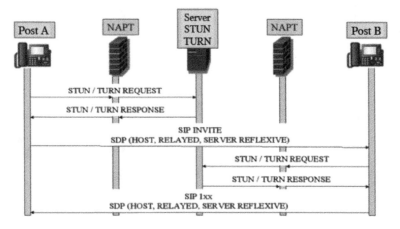

Figure 9.15. *ICE mechanism – inventory phase*

Each post pairs pairwise the local and remote candidate interfaces in order to make candidate pairs. During the connectivity phase, each post tests each candidate pair with the remote post. The connectivity test between the two posts is administered via the exchange of STUN BINDING REQUEST and BINDING RESPONSE messages (Figure 9.16).

The choice of the pair to be retained is made by post A, which informs post B of its choice in a STUN BINDING REQUEST message.

Post A then transmits a reINVITE request, the associated SDP message of which contains the IP address and port number of the pair chosen in order to inform the intermediary entities controlling RTP flow (opening of firewall and service quality level to be applied). Likewise, in its 200 OK response, post B indicates the parameters of the RTP flow.

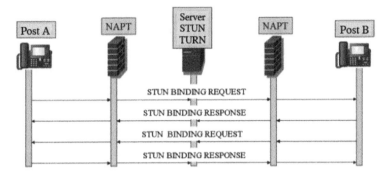

Figure 9.16. *ICE procedure – connectivity checking*

9.2.4. *FTP protocol*

The FTP protocol defines two types of flows: control flow, which enables the passing of commands (access control and parameter-setting for file transfer), and traffic flow, which executes file transfers.

Two operating modes characterize file transfers:

– active mode. The client opens a port number for traffic flow and indicates it to the server in a PORT command. The server initializes the opening of the connection for traffic flow using port 20 as a source port and the port number communicated by the client as a destination port (Figure 9.17);

– passive mode. The client initiates a request to the server to operate in passive mode using the PASV command. In turn, the server indicates to the client the port number it must use as a destination port for traffic flow. In this case, the traffic flow connection is initialized by the client (Figure 9.17).

Passive mode does not pose any problems with regard to crossing the NAT/NAPT device, since both TCP connections are initialized by the client.

Active mode is characterized by the opening of the traffic flow connection by the server. It is imperative that the NAT/NAPT device analyzes the control flow in order to populate the NAT/NAPT table with information concerning traffic flow.

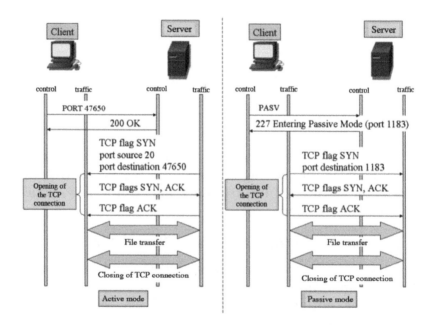

Figure 9.17. *Operating modes of FTP protocol*

Note that the traffic flow can only cross the packet filter if this device also analyzes the control flow and executes a dynamic opening of the traffic flow's IP address and port number. This operation must be completed for both operating modes.

9.2.5. *Fragmentation*

IP packet fragmentation occurs when the packet size is greater than the value of the Maximum Transmission Unit (MTU). If the IP header has encapsulated a TCP or UDP segment, the port numbers appear only in the first fragment.

All the fragments of one single IP packet are identified by the *Identification* field of the IPv4 header or the *Fragmentation* extension of the IPv6 header. This field is used by the destination to reconstitute the source IP packet.

Two different sources can simultaneously send fragments to the same destination. It is possible for these two sources to use the same *Identification* field value.

It is imperative for the NAT/NAPT device to translate the *Identification* field value in order to avoid any confusion on the part of the destination (Figure 9.18).

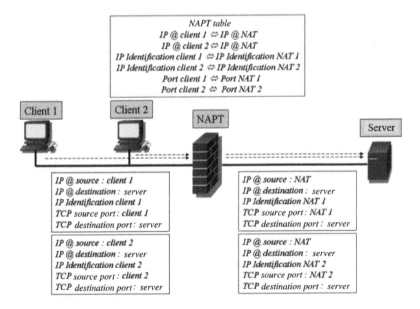

Figure 9.18. *Processing of IP fragments*

Intrusion Detection

10.1. Typology of attacks

Attacks can affect server applications, client applications and network equipment (routers, switches) applications, which are all potential targets. Attacks can also manipulate communications between these entities.

We distinguish five typologies of attacks against a communication between a single source and a single destination (unicast flow) or multiple destinations (multicast or broadcast flow):

– interruption. This attack by Denial of Service (DoS) is intended to block equipment functioning. It consists of flooding the equipment with data or exploiting protocol flaws (Figure 10.1).

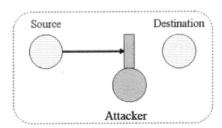

Figure 10.1. *Attack by interruption*

– capture. This attack targets data confidentiality. It is used to access data exchanged between the source and the destination. Capture results from passive network listening or from the insertion of the attacker ("man in the middle" attack) through modification of addressing (Address Resolution Protocol (ARP) or Domain Name System (DNS) attack) between the source and the destination (Figure 10.2);

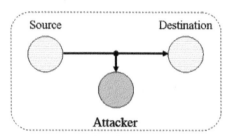

Figure 10.2. *Attack by capture*

– modification. This attack targets data integrity and confidentiality. It occurs when the attacker inserts itself into a communication between the source and the destination, and intercepts and modifies the data in the flow exchanged (Figure 10.3);

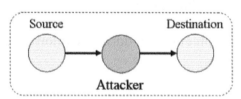

Figure 10.3. *Attack by modification*

– usurpation. This attack enables the attacker to pass itself off as the source and establish communication with the destination (Figure 10.4);

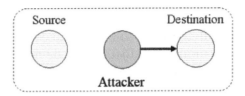

Figure 10.4. *Attack by usurpation*

– injection. This attack is used to insert data into the flow exchanged between the source and the destination (Figure 10.5).

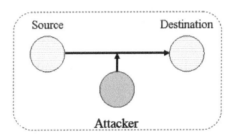

Figure 10.5. *Attack by injection*

Active attacks can be direct when the attacker connects itself directly to its target, by rebound if the attacker uses a relay, or blindly if it cannot see the results of its intervention.

10.2. Methods of detection

To detect the attacks to which different types of equipment are susceptible, Intrusion Detection System (IDS) or Intrusion Prevention System (IPS) devices are required. These two types of device are lumped together under the term Intrusion Detection Prevention System (IDPS).

IDPS devices use numerous methods to detect events: signature-based detection, anomaly-based detection, and stateful protocol analysis. Most IDPS devices use multiple detection methods to ensure broader and more precise detection.

10.2.1. *Signature-based detection*

A signature is a pattern that corresponds to a known threat. Signature-based detection is the process of searching for a string of characters in the data flow, such as:

– an attempt to establish a TELNET session with a root user name, which can constitute a violation of the security policy;

– an email with a file attachment containing the extension .exe, which may contain malicious software.

Signature-based detection is highly effective against known threats, but largely unable to detect previously unknown threats or threats concealed by the use of evasion techniques.

Signature-based detection has a reduced understanding of protocols and is unable to monitor and understand the status of complex communications. This limitation hinders its detection of attacks involving multiple events if none of these events contains a clear indication of an attack.

10.2.2. *Anomaly-based detection*

Anomaly-based detection is the process of comparing activities considered normal to incidents in order to identify significant discrepancies. Profiles can be defined for numerous behavioral attributes, such as the number of emails sent by a user; the number of unsuccessful connection attempts for a host, or the level of processor use for a host in a given period of time.

The principal advantage of anomaly-based detection methods is that they can be very effective at detecting previously unknown threats; for example, when a computer is infected by a new type of malicious software that can consume the computer's processing resources, send a large number of emails, or initiate a large number of connections.

An initial profile is generated for a training period. Profiles can be static or dynamic. Once generated, a static profile does not change. Since normal behavior can change, a static profile will eventually become inaccurate and must be reconfigured periodically. A dynamic profile is continually adjusted in accordance with additional events observed. A dynamic profile is sensitive to evasion attempts. For example, an attacker may slowly increase the frequency and quantity of activity and the IDPS device might consider this normal behavior, and include it in its profile.

Anomaly-based detection often produces false positives if an activity diverges slightly from the profile. For example, if a file transfer takes place once per month, it may not be observed during the training phase.

It is sometimes difficult to determine the cause of a specific report and to confirm whether it is really an alert and not a false positive. This is due to the large number of events and their complexity.

10.2.3. *Protocol analysis*

Stateful protocol analysis specifies how protocols should and should not be used. Stateful protocol means that it is possible to monitor the status of transport and application protocols. For example, when a user initiates a File Transfer Protocol (FTP) file transfer session, the session initially has an unauthenticated status. Unauthenticated users should make only a few commands in this case, such as the display of helpful information or the provision of a username and password.

Stateful protocol analysis uses models defined, for example, by the Request For Comments (RFC) norms published by the standardization organization Internet Engineering Task Force (IETF). Conversely, many norms are not exhaustive, which causes variations between implementations. Moreover, many providers add proprietary functionalities. If the protocols are revised or if the providers modify their implementations, the models of the IDPS device must be updated to reflect these changes.

The main disadvantage of stateful protocol analyses concerns overload caused by session status monitoring. In addition, these methods cannot detect attacks that do not violate the characteristics of the protocol, such as the execution of numerous actions in a short period of time in order to cause a denial of service.

10.3. Technologies

Different types of IDPS device are deployed in the network according to location or function fulfilled:

– Network-based IDPS (N-IDPS): this device enables data monitoring on different segments of a Local Area Network (LAN) and is generally applied to Ethernet interfaces;

– Wireless IDPS (WIDPS): this device enables the surveillance of data being transmitted by wireless-fidelity

(Wi-Fi) radio interface. It constitutes a specific case of the N-IDPS device;

– Home-based IDPS (H-IDPS): this device presents functionalities similar to the devices above and enables data surveillance of hosts only;

– Network Behavior Analysis (NBA): this device is used specifically for traffic analysis in order to detect unusual activity.

One characteristic of IDPS devices is that they cannot provide completely precise detection. When a device identifies a benign activity as being malicious, a false positive occurs. When a device is unable to identify a malicious activity, a false negative occurs. It is difficult to eliminate all false positives and negatives, and in most cases the reduced occurrence of one increases the occurrences of the other.

The Unified Threat Management (UTM) device is used to provide an integrated management of threats including firewall and intrusion detection or prevention functions.

10.3.1. *N-IDPS device*

The N-IDPS device is composed of sensors that monitor LAN network activity on one or more segments. The interfaces monitoring the network accept all incoming packets regardless of their final destination. These sensors are available in two formats:

– instrumental sensors are composed of specialized hardware and software. The hardware is generally optimized for the functions to be executed, and all or part of the software is contained in microprograms for increased effectiveness. N-IDPS devices often use a reinforced operating system that network administrators cannot access directly;

– the N-IDPS function can be obtained in the form of software installed on hardware that is compliant with certain specifications, and which can use a standard operating system.

The sensor is considered active when it is deployed so that the traffic flow must cross it. The principal reason for this is to block attacks. This type of sensor may also include a packet-filtering function.

If a specific protocol is used inappropriately, as with a DoS attack, the active sensor can limit the bandwidth that the protocol can use. This prevents negative repercussions for other sessions in terms of resources.

The active sensor may clean up a packet, which means that the malicious content is replaced by benign content. This can involve the repackaging of Internet Protocol (IP), Transmission Control Protocol (TCP), or User Datagram Protocol (UDP) headers or the removal of infected attachments from emails.

A sensor is considered passive when it is deployed so as to monitor a copy of flow that is not effectively crossing the sensor. The capture of traffic is done via the intermediary of Ethernet switches possessing a mirror interface toward which the traffic can be broadcast. This type of sensor does not cause cuts when an attack is detected.

These sensors are usually installed at key points of the network, such as interfaces between blocks of a LAN network or in the Demilitarized Zone (DMZ) of the access block to the Wide Area Network (WAN). They generally connect to the distribution switch of the various blocks composing the LAN network (Figure 10.6).

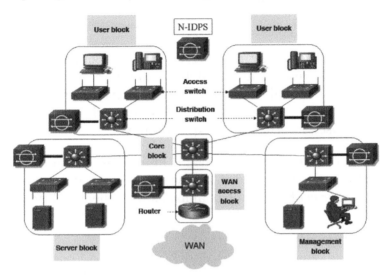

Figure 10.6. *Positioning of N-IDPS sensors*

Sensors generally record data linked to events detected. This data can be used to confirm the validity of alerts, investigate incidents, and correlate events produced by multiple sensors. The following information is frequently used:

– time stamping (usually the date and time at which the event has occurred);

– type of event or alert;

– source and destination IP addresses;

– source and destination TCP or UDP port numbers;

– type and code of Internet Control Message Protocol (ICMP) message;

– decoding of the application layer protocol, such as requests and responses;

– number of bytes transmitted during the session;

– preventive action taken if applicable;

– packet capture.

Many sensors may reconfigure network equipment such as firewalls, routers and switches to block certain types of activity. This makes it possible to prevent an external attacker from penetrating the LAN network, or to put a compromised internal host in quarantine.

Sensors can have significant limitations. They cannot detect attacks contained in encrypted packets and are often unable to execute a complete analysis when heavily loaded.

10.3.2. *WIDPS device*

WIDPS sensors perform the same role as N-IDPS sensors. They monitor traffic on Wireless LAN (WLAN) networks and analyze protocols to identify suspicious activity. However, they function very differently due to the complexity of monitoring wireless communications.

The WIDPS sensor can be a dedicated device that listens passively to traffic. This sensor is linked to the Ethernet switch in order to forward the information it collects to the security management block (Figure 10.7).

The sensor may be an additional function of the Access Point (AP), which must, however, divide its time between supplying traffic to the radio interface and monitoring multiple channels. This solution is better suited to a network functioning on only one radio channel.

The location of WIDPS sensors must also enable the monitoring of zones where no wireless communication should occur.

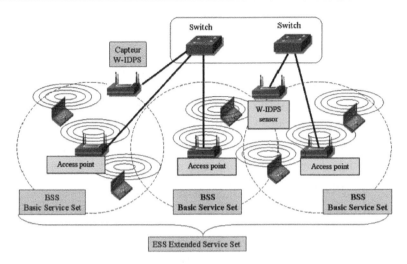

Figure 10.7. *Positioning of WIDPS sensors*

The sensor maintains an inventory of stations observed, access points and users. This inventory is based on BSSID identifiers concerning the Basic Service Set (BSS) and SSID cells involved in the Extended Service Set (ESS) network, as well as MAC addresses. It can also be used to detect the appearance of a new peripheral device or the removal of an existing device.

The sensor can detect attacks, configuration errors and security policy violations in the Wi-Fi physical and data link layers. It does not examine communications in the higher levels. The main events reported by the sensor are as follows:

– the sensor can detect access points, users, and unauthorized WLAN networks both in infrastructure mode and independent mode;

– the sensor can identify access points and users that are not using the appropriate security mechanisms. This is done, for example, via detecting the use of the Wired Equivalent

Privacy (WEP) mechanism instead of the Wi-Fi Protected Access (WPA) mechanism;

– the sensor can detect the presence of active scanners that generate traffic to identify unsecured or weakly protected WLAN networks. However, they cannot detect the use of passive sensors that analyze observed traffic;

– the sensor can detect anomalies corresponding to a higher-than-normal use of an access point, for example activity during off-peak hours;

– the sensor can detect physical attacks that interfere with radio transmission.

The physical location of the threat detected is determined via triangulation, which results from the estimation by multiple sensors of the approximate distance to the threat, calculated from the signal received.

Sensors can also have intrusion prevention capabilities. Some sensors can request that an Ethernet switch block WLAN network activity, but this method cannot stop malicious actions on the WLAN network.

10.3.3. *H-IDPS device*

H-IDPS sensors monitor the characteristics of a single host and the events that occur within this host for any suspicious activity.

The sensor is generally deployed for important hosts such as servers containing sensitive information, located in the server block, the management block, and the demilitarized zone of the WAN network access block (Figure 10.8).

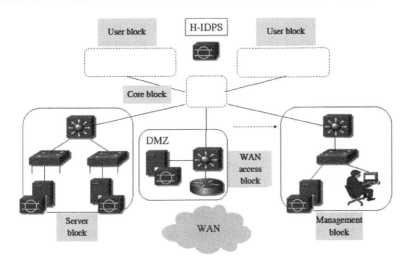

Figure 10.8. *Positioning of H-IDPS sensor*

Most sensors modify the internal architecture of the hosts in which they are installed using code placed between the existing layers. A less intrusive device reduces detection capabilities.

Security functions include code analysis, traffic analysis and filtering, file system surveillance, event log analysis, and monitoring of application configurations or network parameters.

Some sensors offer additional capabilities related to intrusion detection and prevention, such as restrictions on the use of removable supports, or detection of the use of audiovisual devices.

10.3.4. *NBA device*

NBA devices examine traffic or traffic statistics to identify unusual flows, some forms of malicious software, and violations of security policy.

The NBA device is composed of sensors, some of which are similar to those of the N-IDPS device in that they examine packets to monitor the activity of various LAN network blocks.

Other types of sensors do not monitor segments directly, but rely on the flows provided by network equipment. NetFlow, sFlow and IP-Fix are formats defined for this type of data transfer. These flows can be generated by distribution switches (Figure 10.9).

The NBA device has vast information collection capacities due to the fact that knowledge of host characteristics is necessary for most detection techniques. The sensors can automatically create and maintain lists of hosts; monitor the use of a port; create passive digests; and collect detailed information about hosts.

The following types of events can be detected by the sensors:

– Denial of Service (DoS). These attacks generally involve a significant increase in the bandwidth used or a larger number of packets emitted by or sent to a host;

– scanning. This attack can be detected by flow models pertaining to the application layer, the transport layer (for example the scanning of TCP and UDP port numbers), and the network layer (for example the scanning of IP addresses);

– worms. This attack may be detected indirectly, as worms can initiate unusual communications between hosts using atypical port numbers;

– violation of security policy. Sensors specify detailed policies such as contacts authorized for a group of hosts or types of activities permitted during certain hours or days of the week. They detect the appearance of new hosts or new unauthorized applications.

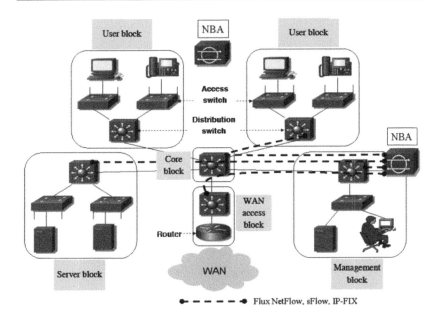

Figure 10.9. *Positioning of NBA sensor*

Bibliography

[ABO 03] ABOBA B., CALHOUN P., "RADIUS (Remote Authentication Dial in User Service) support for Extensible Authentication Protocol (EAP)", *RFC 3579*, September 2003.

[ABO 04] ABOBA B., BLUNK L., VOLLBRECHT J., *et al.*, "Extensible authentication protocol (EAP)", *RFC 3748*, June 2004.

[AGG 06] AGGARWAL R., TOWNSLEY M., DOS SANTOS M., "Transport of Ethernet frames over Layer 2 Tunneling Protocol version 3 (L2TPv3)", *RFC 4719*, November 2006.

[AND 07] ANDERSSON L., MINEI I., THOMAS B., "LDP Specification", *RFC 5036*, October 2007.

[BAR 12] BARKER W.C., BARKER E., "Recommendation for the Triple Data Encryption Algorithm (TDEA) Block Cipher", *NIST 800-67*, January 2012.

[BLU 02] BLUMENTHAL U., WIJNEN B., "User-based Security Model (USM) for version 3 of the Simple Network Management Protocol (SNMPv3)", *RFC 3414*, December 2002.

[CHE 05] CHERNICK C.M., EDINGTON III C., FANTO M.J., *et al.*, "Guidelines for the selection and use of Transport Layer Security (TLS) implementations", *NIST 800-52*, June 2005.

[DEE 98] DEERING S., HINDEN R., "Internet Protocol, version 6 (IPv6) specification", *RFC 2460*, December 1998.

[DIE 08] DIERKS T., RESCORLA E., "The Transport Layer Security (TLS) protocol version 1.2", *RFC 5246*, August 2008.

[FRA 05] FRANKEL S., KENT K., LEWKOWSKI R., et al., "Guide to IPsec VPNs", *NIST 800-77*, December 2005.

[FRA 07] FRANKEL S., EYDT B., OWENS L., et al., "Establishing wireless robust security networks: a guide to IEEE 802.11i", *NIST 800-97*, February 2007.

[FU 07] FU D., SOLINAS J., "ECP groups for IKE and IKEv2", *RFC 4753*, January 2007.

[FUN 08] FUNK P., BLAKE-WILSON S., "Extensible Authentication Protocol Tunneled Transport Layer Security Authenticated Protocol Version 0 (EAP-TTLSv0)", *RFC 5281*, August 2008.

[HAR 02] HARRINGTON D., PRESUHN R., WIJNEN B., "Message processing and dispatching for the Simple Network Management Protocol (SNMP)", *RFC 3412*, December 2002.

[HOL 01] HOLDREGE M., SRISURESH P., "Protocol complications with the IP Network Address Translator", *RFC 3027*, January 2001.

[HUT 05] HUTTUNEN A., SWANDER B., VOLPE V., et al., "UDP encapsulation of IPsec ESP packets", *RFC 3948*, January 2005.

[KAU 10] KAUFMAN C., HOFFMAN P., NIR Y., et al., "Internet Key Exchange protocol version 2 (IKEv2)", *RFC 5996*, September 2010.

[KEN 05a] KENT S., "IP authentication header", *RFC 4302*, December 2005.

[KEN 05b] KENT S., "IP Encapsulating Security Payload (ESP)", *RFC 4303*, December 2005.

[KEN 05c] KENT S., SEO K., "Security architecture for the Internet Protocol", *RFC 4301*, December 2005.

[KIV 03] KIVINEN T., KOJO M., "More Modular Exponential (MODP) Diffie–Hellman groups for Internet Key Exchange (IKE)", *RFC 3526*, May 2003.

[KIV 05] KIVINEN T., SWANDER B., HUTTUNEN A., et al., "Negotiation of NAT-traversal in the IKE", *RFC 3947*, January 2005.

[KRA 97] KRAWCZYK H., BELLARE M., CANETTI R., "HMAC: keyed-hashing for message authentication", *RFC 2104*, February 1997.

[LAU 05] LAU J., TOWNSLEY M., GOYRET I., "Layer Two Tunneling Protocol – version 3 (L2TPv3)", *RFC 3931*, March 2005.

[LE 02] LE FAUCHEUR F., WU L., DAVIE B., *et al.*, "Multiprotocol Label Switching (MPLS) support of differentiated services", *RFC 3270*, May 2002.

[LEP 08] LEPINSKI M., KENT S., "Additional Diffie–Hellman groups for use with IETF standards", *RFC 5114*, January 2008.

[LEV 02] LEVI D., MEYER P., STEWART B., "Simple Network Management Protocol (SNMP) applications", *RFC 3413*, December 2002.

[MAH 10] MAHY R., MATTHEWS P., ROSENBERG J., "Traversal Using Relays around NAT (TURN): relay extensions to Session Traversal Utilities for NAT (STUN)", *RFC 5766*, April 2010.

[MAR 06a] MARTINI L., ROSEN E., EL-AAWAR N., *et al.*, "Encapsulation methods for transport of Ethernet over MPLS networks", *RFC 4448*, April 2006.

[MAR 06b] MARTINI L., ROSEN E., EL-AAWAR N., *et al.*, "Pseudowire setup and maintenance using the Label Distribution Protocol (LDP)", *RFC 4447*, April 2006.

[MCG 10] MCGREW D., RESCORLA E., "Datagram Transport Layer Security (DTLS) extension to establish keys for the Secure Real-time Transport Protocol (SRTP)", *RFC 5764*, May 2010.

[NIS 01] NIST, "Advanced Encryption Standard (AES)", *NIST FIPS PUB 197*, November 2001.

[NIS 08] NIST, "The keyed-hash message authentication code (HMAC)", *NIST FIPS PUB 198-1*, July 2008.

[NIS 12] NIST, "Secure Hash Standard (SHS)", *NIST FIPS PUB 180-4*, March 2012.

[PHE 08] PHELAN T., "Datagram Transport Layer Security (DTLS) over the Datagram Congestion Control Protocol (DCCP)", *RFC 5238*, May 2008.

[POS 81] POSTEL J., "Internet Protocol", *RFC 791*, September 1981.

[RES 99] RESCORLA E., "Diffie–Hellman key agreement method", *RFC 2631*, June 1999.

[RES 12] RESCORLA E., MODADUGU N., "Datagram Transport Layer Security version 1.2", *RFC 6347*, January 2012.

[RIG 00] RIGNEY C., WILLENS S., RUBENS A., "Remote Authentication Dial in User Service (RADIUS)", *RFC 2865*, June 2000.

[RIV 92] RIVEST R., *"The MD5 message-digest algorithm"*, RFC 1321, April 1992.

[ROS 01] ROSEN E., VISWANATHAN A., CALLON R., "Multiprotocol Label Switching Architecture", *RFC 3031*, January 2001.

[ROS 06] ROSEN E., REKHTER Y., "BGP/MPLS IP Virtual Private Networks (VPNs)", *RFC 4364*, February 2006.

[ROS 08] ROSENBERG J., MAHY R., MATTHEWS P., *et al.*, "Session Traversal Utilities for NAT (STUN)", *RFC 5389*, October 2008.

[ROS 10] ROSENBERG J., "Interactive Connectivity Establishment (ICE): a protocol for Network Address Translator (NAT) traversal for Offer/Answer Protocols", *RFC 5245*, April 2010.

[SCA 07] SCARFONE K., MELL P., "Guide to Intrusion Detection and Prevention Systems (IDPS)", *NIST 800-94*, February 2007.

[SCA 08] SCARFONE K., HOFFMAN P., "Guidelines on firewalls and firewall policy", *NIST 800-41*, July 2008.

[SIM 08] SIMON D., ABOBA B., HURST R., "The EAP-TLS authentication protocol", *RFC 5216*, March 2008.

[SRI 99] SRISURESH P., HOLDREGE M., "IP Network Address Translator (NAT) terminology and considerations", *RFC 2663*, August 1999.

[SRI 01] SRISURESH P., EGEVANG K., "Traditional IP Network Address Translator (Traditional NAT)", *RFC 3022*, January 2001.

[SRI 09] SRISURESH P., FORD B., SIVAKUMAR S., *et al.*, "NAT behavioral requirements for ICMP", *RFC 5508*, April 2009.

[TUE 11] TUEXEN M., SEGGELMANN R., RESCORLA E., "Datagram Transport Layer Security (DTLS) for Stream Control Transmission Protocol (SCTP)", *RFC 6083*, January 2011.

[WIJ 02] WIJNEN B., PRESUHN R., MCCLOGHRIE K., "View-based Access Control Model (VACM) for the Simple Network Management Protocol (SNMP)", *RFC 3415*, December 2002.

[YLO 06a] YLONEN T., LONVICK C., "The Secure Shell (SSH) authentication protocol architecture", *RFC 4252*, January 2006.

[YLO 06b] YLONEN T., LONVICK C., "The Secure Shell (SSH) connection protocol", *RFC 4254*, January 2006.

[YLO 06c] YLONEN T., LONVICK C., "The Secure Shell (SSH) protocol architecture", *RFC 4251*, January 2006.

[YLO 06d] YLONEN T., LONVICK C., "The Secure Shell (SSH) Transport Layer Protocol", *RFC 4253*, January 2006.

Standards

"Port-based Network Access Control", *IEEE 802.1X*, February 2010.

"Medium Access Control (MAC) Security Enhancements", *IEEE 802.11i*, July 2004.

"Virtual bridged local area networks", *IEEE 802.1Q*, May 2006.

"Virtual bridged local area networks – amendment 4: provider bridges", *IEEE 802.1ad*, May 2006.

Index

Other titles from

in

Networks and Telecommunications

2014

VAN METER Rodney
Quantum Networking

XIONG Kaiqi
Resource Optimization and Security for Cloud Services

2013

ASSING Dominique, CALÉ Stéphane
Mobile Access Safety: Beyond BYOD

BEN MAHMOUD Mohamed Slim, LARRIEU Nicolas, PIROVANO Alain
Risk Propagation Assessment for Network Security: Application to Airport Communication Network Design

BEYLOT André-Luc, LABIOD Houda
Vehicular Networks: Models and Algorithms

BRITO Gabriel M., VELLOSO Pedro Braconnot, MORAES Igor M.
Information-Centric Networks: A New Paradigm for the Internet

BERTIN Emmanuel, CRESPI Noël
Architecture and Governance for Communication Services

DEUFF Dominique, COSQUER Mathilde
User-Centered Agile Method

DUARTE Otto Carlos, PUJOLLE Guy
Virtual Networks: Pluralistic Approach for the Next Generation of Internet

FOWLER Scott A., MELLOUK Abdelhamid, YAMADA Naomi
LTE-Advanced DRX Mechanism for Power Saving

JOBERT Sébastien *et al.*
Synchronous Ethernet and IEEE 1588 in Telecoms: Next Generation Synchronization Networks

MELLOUK Abdelhamid, HOCEINI Said, TRAN Hai Anh
Quality-of-Experience for Multimedia: Application to Content Delivery Network Architecture

NAIT-SIDI-MOH Ahmed, BAKHOUYA Mohamed, GABER Jaafar,
WACK Maxime
Geopositioning and Mobility

PEREZ André
Voice over LTE: EPS and IMS Networks

2012

AL AGHA Khaldoun
Network Coding

BOUCHET Olivier
Wireless Optical Communications

DECREUSEFOND Laurent, MOYAL Pascal
Stochastic Modeling and Analysis of Telecoms Networks

DUFOUR Jean-Yves
Intelligent Video Surveillance Systems

EXPOSITO Ernesto
Advanced Transport Protocols: Designing the Next Generation

JUMIRA Oswald, ZEADALLY Sherali
Energy Efficiency in Wireless Networks

KRIEF Francine
Green Networking

PEREZ André
Mobile Networks Architecture

2011

BONALD Thomas, FEUILLET Mathieu
Network Performance Analysis

CARBOU Romain, DIAZ Michel, EXPOSITO Ernesto, ROMAN Rodrigo
Digital Home Networking

CHABANNE Hervé, URIEN Pascal, SUSINI Jean-Ferdinand
RFID and the Internet of Things

LABIOD Houda
Wireless Ad Hoc and Sensor Networks

LECOY Pierre
Fiber-optic Communications

MELLOUK Abdelhamid
*End-to-End Quality of Service Engineering in Next Generation
Heterogeneous Networks*

PAGANI Pascal *et al.*
Ultra-wideband Radio Propagation Channel

2007

BENSLIMANE Abderrahim
Multimedia Multicast on the Internet

PUJOLLE Guy
Management, Control and Evolution of IP Networks

SANCHEZ Javier, THIOUNE Mamadou
UMTS

VIVIER Guillaume
Reconfigurable Mobile Radio Systems

Printed and bound by CPI Group (UK) Ltd, Croydon, CR0 4YY

27/10/2024

14580317-0004